Microautoradiography and Electron Probe Analysis

Their Application to Plant Physiology

W. O. Abel P. Dörmer W. Eschrich E. Fritz
R. G. Herrmann A. Läuchli U. Lüttge J. B. Passioura
J. D. Pickett-Heaps

Edited by U. Lüttge

With 78 Figures

Springer-Verlag
New York · Heidelberg · Berlin 1972

ISBN 0-387-05950-4 Springer-Verlag New York Heidelberg Berlin
ISBN 3-540-05950-4 Springer-Verlag Berlin Heidelberg New York

 Library of Congress Catalog Card Number 72-97599
Typesetting, printing and binding: Konrad Triltsch, Graphischer Betrieb, Würzburg

Contents

Introductory Chapter

Botanical Applications of Microautoradiography

Ulrich Lüttge
Fachbereich Biologie – Botanik – der Technischen Hochschule, Darmstadt, W.-Germany

Historically, microautoradiography has already been a successful tool in medicine and zoology while progress in the application of the technique to physiological problems in the plant sciences was slow. There may be several reasons for this, the most pertinent of which is the different nature of plant cells and animal cells. Although the microautoradiographic technique is applicable in principle to both plant and animal tissues, plant cells with a prominent cell wall and with a large central sap vacuole pose different preparative problems than animal cells. Hence, it is not surprising that, in the classical monograph of microautoradiographic techniques published in 1955 by Boyd, botanical applications are extremely few. And even in more recent publications (Rogers, 1967; Fischer and Werner, 1971) there is very little reference to microautoradiographic work with plant material (in addition, see chapter 16 in Jensen, 1962).

However, the excellent study of ^{35}S and ^{32}P translocation in the phloem by Biddulph (1956) and the investigation of the localisation of lignin-biosynthesis in young branches of spruce by Freudenberg et al. (1955) led the way, and in the past decade microautoradiography has been widely used in plant physiology. In many cases it has provided insights into plant physiological problems, which would not have been possible with other techniques.

A major section of plant physiology in which microautoradiography plays a considerable role is transport physiology. A large number of microautoradiographic studies of long distance transport in the phloem and the xylem have been successfully performed, but problems of short distance transport and cell compartmentation have also been investigated using microautoradiography.

Localisation of biosynthesis of cell components on a tissue level (e.g. lignin, FREUDENBERG et al., 1955) on the cellular level and even on the subcellular level (e.g. biosynthesis of cell wall constituents; see chapter 6) is another major field of application of microautoradiography in the plant sciences.

Investigation of these problems requires preparative techniques for the microautoradiographic detection of labeled compounds both soluble and insoluble in water. They also require techniques for resolution on the tissue level, the cellular level, and the subcellular level. The authors contributing to this volume, report their own experience with particular techniques and have listed the literature they consider pertinent. The aim of this volume, then, is to provide the plant physiologist of an account of successful microautoradiographic applications and of particular pitfalls which have been found. It should thus help the plant physiologist in designing experiments with plant cells and tissues. The 3rd—5th chapters deal with the microautoradiographic detection of water soluble inorganic and ^{3}H- or ^{14}C-organic compounds, and of insoluble compounds on the tissue and cellular levels. The 6th chapter is concerned with autoradiography on the subcellular level.

Autoradiography only provides data on relative or, in some cases, on absolute amounts of radioactivity localised in a given area of a tissue section. For many problems of translocation and of biosynthesis, however, data on specific activities would be more valuable and meaningful. This could be achieved if, in addition to the radioactivity, the chemical quantity of a substance or element present in that particular area of a section was known. In some instances, data of the latter kind may be provided by electron probe analysis. Microautoradiography and electron probe analysis have been separately applied to similar physiological problems, e.g. in the investigation of the role of ion transport in regulation of stomatal aperture (Fig. 1). If used simultaneously, these techniques allow the determination of specific activities. For these reasons, and also because of the growing importance of electron probe analysis in cytology and plant physiology, an account of this technique has been included in the 7th chapter giving particular emphasis to its relations to microautoradiography.

The reader will not find, in this book, discussions of the basic principles of autoradiography and the photographic process, and of the practical problems related to them. These problems apply in all autoradiographic work and have been well covered elsewhere (e.g. BOYD, 1955; ROGERS, 1967; FISCHER and WERNER, 1971).

However, the volume begins with two more general chapters. It was felt, that there should be some advice on applications of methods — other than silver grain counting — for quantitative measurements of microautoradiographs. The most important technique is photometry of microautoradiographs, which in investigations using animal tissues has been considerably

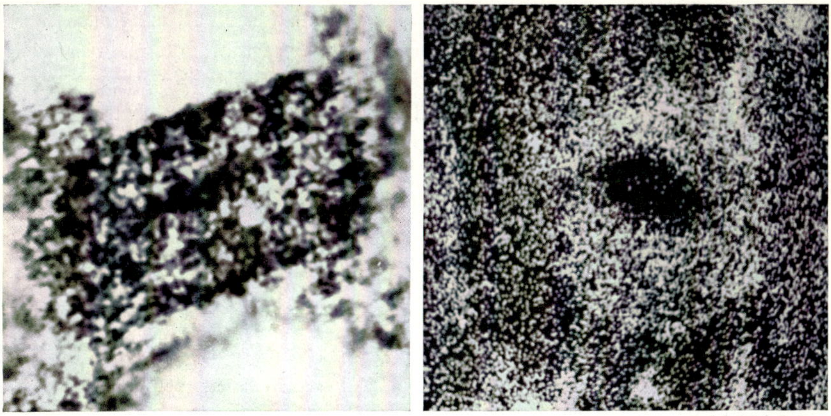

Fig. 1. (a) Microautoradiographic demonstration of the accumulation of $^{35}SO_4^{--}$ in the stomatal guard cells of *Chenopodium album* leaves. (Sulfate uptake for 7 hrs by epidermal strips from a 0.1 mM $K_2{}^{35}SO_4$-solution, 400 mCi ^{35}S/mmole SO_4^{--}, at 25 °C, exposure time 20 days. The microphotograph was taken in the transmitted light so that the developed silver grains appear as black dots.) Cf. OSMOND et al. 1969. Magnification ca. 900 ×. (b) Electron probe analysis-demonstration of K^+-accumulation in the guard cells of an opened stomate of a tobacco leaf illuminated for 1.5 hrs. Magnification ca. 800 ×. With kind permission from SAWHNEY and ZELITCH (1969)

refined in the recent years, and which is evaluated in the 1st chapter. Although the photometric technique has been applied in plant microautoradiography (LÄUCHLI and LÜTTGE, 1968; OSMOND et al., 1969; LÜTTGE et al., 1971) experience based on experiments with plant material is limited. However, in respect to evaluations of photometry of autoradiographs the origin of the tissue, i.e. from plant or animal specimen, does not appear to be highly important. The second chapter presents some original evaluations of quantitative autoradiography in the presence of crossfire, which is important particularly in the use of isotopes with high energy radiation. It is anticipated that these two chapters will encourage the application of modern techniques of quantitative autoradiography and trigger some thinking about mathematical treatments of resolution problems.

Finally, having introduced the book in an optimistic vein, it has to be said, that microautoradiography will often prove to be a difficult technique to apply. It is important to decide at the outset if microautoradiography is the most appropriate approach to a problem or if there are other methods more suited. As illustrated in the following example the decision will often not only depend on the nature of the problem itself but also on subjective predilection of the investigator. To investigate ion accumulation within chloroplasts in intact plant cells the biochemist-physiologist may tend to use chloroplast isolation techniques which minimise redistribution of ions during preparation. The anatomist-cytologist may be more inclined to apply microautoradiography with preparative provisions allowing detection of water soluble material *in situ.* Indeed both approaches have been used in investigations of ionic relations of chloroplasts *in vivo* and have independently suggested that under certain conditions a high percentage of the total ion content of a plant cell is localised within the chloroplasts. (For the example of chloride see Fig. 2 and Table 1.)

Fig. 2. Microautoradiographic demonstration of $^{36}Cl^-$-accumulation in chloroplasts (Ch) of a leaf of the halophyte *Limonium vulgare* (from Ziegler and Lüttge, 1967; uptake for 8 hrs via the petiole of the isolated leaf immersed in a 0.5 M $Na^{36}Cl$ solution, 0.3 mCi/M Cl^-, exposure time 14 days). Magnification 1000 ×. (a) Microphotography in transmitted light (developed silver grains appear as black dots). (b) Microphotography in incident light (developed silver grains appear as reflecting bright dots). By chloroplast isolation and chemical analysis the rather high amount of 1.25 mMoles Cl^- per mg dry chloroplasts has been observed in *Limonium* leaves (Larkum, 1968)

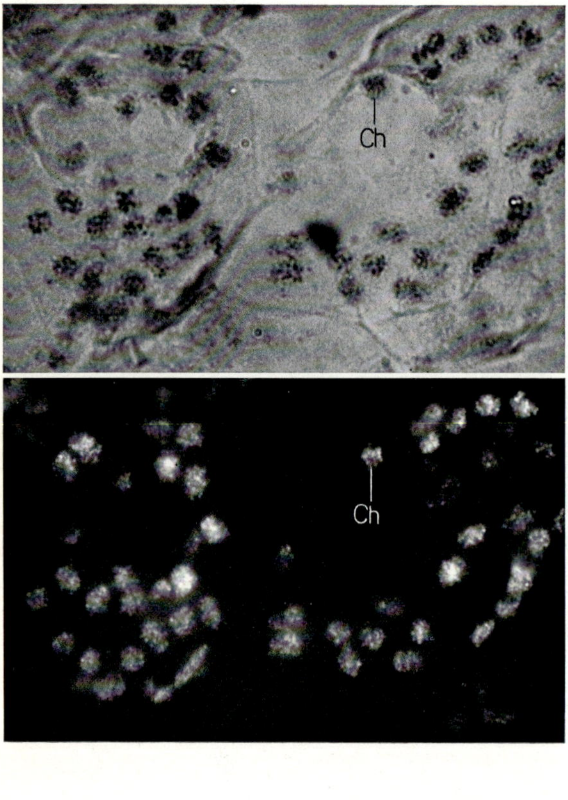

Table 1. Cl^--concentration in various cell compartments of *Nitella flexilis* (Kishimoto and Tazawa, 1965) and *Tolypella intricata* (Larkum, 1968)

	External medium	Cl^--concentration [mmoles/l] Chloroplast-layer	Streaming cytoplasm	Vacuole
Nitella	1.3	136	36	136
Tolypella	1.4	340	23—31	116—136

References

Biddulph, S. F.: Visual indication of ^{35}S and ^{32}P translocation in the phloem. Amer. J. Bot. **43**, 143—148 (1956).

Boyd, G. A.: Autoradiography in biology and medicine. New York. Academic Press 1955.

Fischer, H. A., Werner, G.: Autoradiographie. Walter de Gruyter, Berlin-New York 1971.

Freudenberg, K., Reznik, H., Fuchs, W., Reichert, M.: Untersuchung über die Entstehung des Lignins und des Holzes. Naturwissenschaften **42**, 29—35 (1955).

Jensen, W. A.: Botanical histochemistry. San Francisco-London. Freeman and Co. 1962.

Kishimoto, U., Tazawa, M.: Ionic composition of the cytoplasm of *Nitella flexilis*. Plant and Cell Physiology **6**, 507—518 (1965).

Läuchli, A., Lüttge, U.: Untersuchung der Kinetik der Ionenaufnahme in das Cytoplasma von *Mnium*-Blattzellen mit Hilfe der Mikroautoradiographie und der Röntgen-Mikrosonde. Planta **83**, 80—98 (1968).

Larkum, A. W.: Ionic relations of chloroplasts *in vivo*. Nature (Lond.) **218**, 447—449 (1968).

Lüttge, U., Pallaghy, C. K., Willert, K. von: Microautoradiographic investigations of sulfate uptake by glands and epidermal cells of water lily (*Nymphaea*) leaves with special reference to the effect of poly-L-lysine. J. Membrane Biol. **4**, 395—407 (1971).

Osmond, C. B., Lüttge, U., West, K. R., Pallaghy, C. K., Shacher-Hill, B.: Ion absorption in *Atriplex* leaf tissue. II. Secretion of ions to epidermal bladders. Aust. J. Biol. Sci. **22**, 797—814 (1969).

Rogers, A. W.: Techniques of autoradiography. Elsevier, Amsterdam 1967.

Sawhney, B. L., Zelitch, I.: Direct determination of potassium ion accumulation in guard cells in relation to stomatal opening in light. Plant Physiol. **44**, 1350—1354 (1969).

Ziegler, H., Lüttge, U.: Die Salzdrüsen von *Limonium vulgare* II. Die Lokalisierung des Chlorids. Planta **74**, 1—17, (1967).

Chapter 1

Photometric Methods in Quantitative Autoradiography*

P. DÖRMER
Institute of Hematology of the Gesellschaft für Strahlen- und Umweltforschung,
in association with EURATOM, München, W.-Germany

Contents

* The investigations were performed under the association contract EURATOM — GSF No. 031-641 BIAD for hematology.

1.1. *Introduction*

In biological literature, several definitions of quantitative autoradiography are given. The term is defined as either the determination and comparison of the density of silver grains above various structures or under varying conditions, or the determination of absolute quantities of radioactivity. In both these cases, photometric measurement serves for more rapid and more exact evaluation of grain densities than would be possible by visual counting of the grains.

The equipment generally used for the photometric measurement of silver grains consists of a microscope, a photocell, an electronic amplifier system and a display unit. Grains can be made accessible to photometric evaluation by various kinds of microscopic illumination:

1. Substage bright-field illumination.
2. Substage dark-field illumination.
3. Incident dark-field illumination.
4. Vertical bright-field illumination.

With all these types of illumination, the relationship between the luminous flux I absorbed by the film, scattered into the objective and reflected or diffracted, and the flux I_0 which is not affected by the film is used as a measure of grain density. Since these are differential measurements, the light beam I transmitted by the film is in itself a measure of grain density if the luminous flux I_0 incident on the grains is kept constant. This approach has been used in a large number of measuring setups.

Since the diameter of silver grains is of the order of the wavelengths used for photometric measurement and since chemically developed grains have a completely irregular surface structure, terms like reflection and absorption become questionable if used in the sense of physics. If these terms nevertheless are used in the following, it is only to facilitate explanation of the principles of special interest to the biologist.

1.2. *Optical Principles of Silver-Grain Photometry*

1.2.1. Photometry by Substage Bright-Field Illumination

Parallel light rays incident on a photographic emulsion are not only absorbed and reflected on passing through it, but also scattered. As a result,

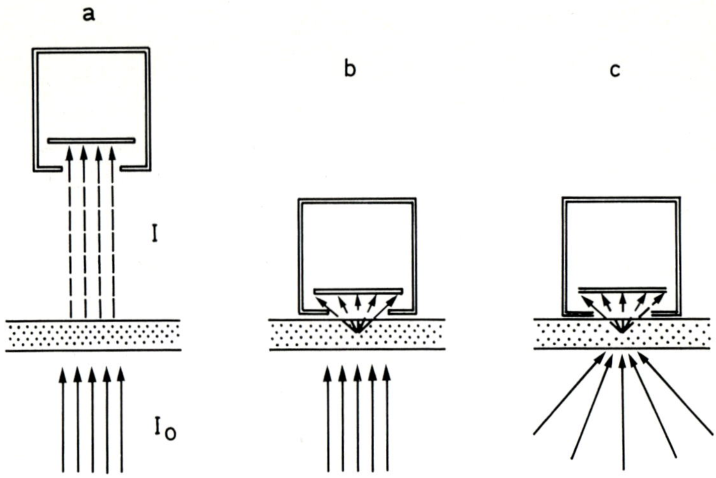

Fig. 1.1. Optical diagram for vertical bright-field illumination. In relation to the luminous flux (I_0) incident on the object, a different flux I is measured when determining a) the specular, b) the diffuse and c) the doubly diffuse blackening

part of the light emerges from the emulsion at an angle different from its angle of incidence. As a measure of light absorption we may record the luminous flux that has neither been absorbed, reflected nor scattered by the photographic emulsion. This type of light measurement is called specular (Fig. 1.1a). On the other hand, the light passing through the emulsion can also be recorded including the scattered component. In this case we speak of diffuse-light measurement (Fig. 1.1b). If the light incident on the emulsion is diffused and the scattered component is also measured, then doubly diffuse density is recorded (Fig. 1.1c).

It is obvious that the luminous flux recorded in relation to the flux incident on the emulsion will vary for a given emulsion and grain density, depending on the measuring technique used. The blackening D (transmission density) of a photographic emulsion is defined as $\log_{10} \frac{I_0}{I}$ when the transmitted light is diffused as shown in Fig. 1.1b. I_0 is the value measured for the light incident on the emulsion and I the value measured after passage through the emulsion. Transmission is defined as $T = \frac{I}{I_0}$ and absorption as $A = 1 - T = 1 - \frac{I}{I_0}$.

Most macrodensitometers work on the principle of diffuse densitometry. Due to the properties of the objectives used, however, the great majority of microdensitometers also register part of the light scattered by the silver grains, although they are unable to measure all the scattered light. We have here an intermediate stage between registration of specular and diffuse blackening. ODEBLAD (1952) compared the results obtained with a microdensitometer developed in his laboratory with those of a commercial macrodensitometer. He found that the relationship between the blackening determined in the two photometers varied considerably with the grain density.

Diffuse blackening is related to specular blackening by a relationship known as Callier's Q factor. This depends, among other things, on the wavelength used for measurement, on the mean grain size and the gradation of the photographic emulsion (EGGERT and KUESTER, 1938; HERZ, 1969; JAMES and HIGGINS, 1948). This means that it can also be influenced by the processing conditions. Unless reference densities can be provided on the autoradiographs, it is therefore advisable when working with substage bright-field illumination or other illuminating techniques to insure constant optical settings of the photometer on the one hand and constant processing conditions of the nuclear emulsions on the other.

The measurement of autoradiographic grain densities in substage bright-field illuminations has the great disadvantage that the radioactive specimen can remain in contact with the nuclear emulsion only if a gross measuring error is tolerated. This is due to the fact that even an unstained tissue section may absorb more light than the silver grains above it. Transmitted-light bright-field photometry is therefore generally unsuitable if correlation of the grains with the microscopic structures is required at higher powers (PELC, 1959).

1.2.1.1. Measurement of Absorbed Light

In 1968, FISCHER and WERNER described a single-beam microscope photometer for measuring the light absorbed by nuclear emulsions separated from the radioactive specimens before processing. Measuring absorption instead of blackening involves fewer technical problems in single-beam photometers. The authors show that in the range of low grain densities there is sufficient linearity between the number of grains and the absorbed luminous flux. The use of dry objectives of various powers did not reveal any difference in absorption measurements.

1.2.1.2. The Flying Spot Principle

As early as 1953 DUDLEY and PELC described a flying spot microscope suited for counting grains in nuclear emulsions. The flying spot principle uses a light beam which a television tube mounted above the microscope emits downwards in the direction of the condenser. The microscope objective focuses the beam on the autoradiograph, which is scanned in a raster pattern. The absorption occurring as the beam crosses a silver grain is counted as a pulse by a recording unit situated near the condenser. Although this principle of automatic grain counting is a very neat solution, it does have a few disadvantages. The greatest drawback is undoubtedly the high instrument outlay. Another disadvantage is the fact that separate recording of two closely adjacent grains is extremely difficult since they will be recorded as a single pulse. This type of photometer is therefore only suited for very low grain densities. The authors indicate a linear relationship between the number of beta particles emitted into the nuclear emulsion and the recorded silver grains up to a density of 40 grains per 100 μm^2.

1.2.1.3. Microdensitometry

Microdensitometry is the photometric procedure most frequently used in quantitative autoradiography (ALTMAN, 1963a, 1963b; ANDRESEN et al., 1953; AXELROD and HAMILTON, 1947; BERLIN and ULLBERG, 1963; BLACKETT, KEMBER and LAMERTON, 1959; BOSTRÖM, ODEBLAD and FRIBERG, 1952; BRANSON and HANSBOROUGH, 1948; DUDLEY and DOBYNS, 1949; FICQ and ERRERA, 1958; KUTZIM, 1963; MARSHALL, ROWLAND and JOWSEY, 1959; MAZIA, PLAUT and ELLIS, 1955; ODEBLAD, 1952; VAUGHAN and OWEN, 1959; WASER and LÜTHI, 1962). ALTMAN (1963a) investigated the relationship between blackening and the number of grains in two different nuclear emulsions and found rigorous linearity in the range of visually countable grain densities. ODEBLAD (1952) gave a detailed description of a microdensitometer which was designed by him. The microdensitometer used by WASER and LÜTHI (1962) is a double-beam instrument in which the absorption in one beam is compared with that in the other by means of a sliding neutral wedge. MARSHALL, ROWLAND and JOWSEY (1959) discussed the theoretical problems of quantitative autoradiography with special reference to a simplification of microdensitometric measurement.

The major part of this microdensitometric work was performed on autoradiographs produced by the contact method (for procedure, see EVANS, 1948; GROSS et al., 1951). For theoretical reasons (see 1.3.), even today preference over other optical principles of photometric evaluation of autoradiographs must be given to densitometry. Unfortunately, however, the usefulness of this method is rather limited because the radioactive specimen and the nuclear emulsion have to be separated for measurement.

In the process described by MAZIA, PLAUT and ELLIS (1955), the grain distribution in the autoradiograph is first transformed into a one-dimensional intensity distribution by photographing it through a cylindrical lens.

The one-dimensional distribution is then quantified with the aid of a densitometer. The major disadvantage of this method is the additional photographic transformation of the autoradiograph.

Another special form of densitometry is the method of photographic colorimetry described by JOWSEY (1966) and JOWSEY and co-workers (1965). In this case, a photographic print of bluish-white tone is made of the autoradiograph, in which a deposit of water-insoluble blue dye forms during development instead of silver grains. This dye is extracted from the sites to be examined and measured in a spectral photometer. The authors show that within a narrow density range of 0 to 0.02 there is a linear relationship between the quantity of blue dye deposited and the radioactive dose to which the autoradiograph was exposed.

1.2.2. Photometry by Substage Dark-Field Illumination

In dark-field work, the illuminating beam is directed so that the light can be captured by the objective only if it is diffracted by the specimen (Fig. 1.2). This is the case above all at the surface of the silver grains which therefore stand out brightly against the dark background. For substage

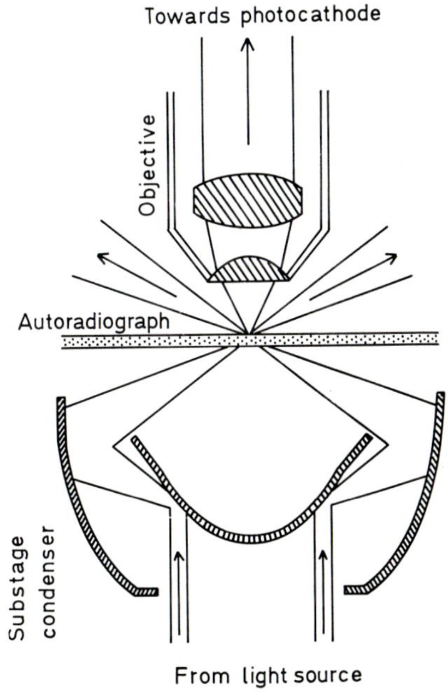

Fig. 1.2. Optical diagram for substage dark-field illumination. The lower limit of the illuminating aperture of the substage condenser is larger than the viewing aperture of the objective. As a result, only light diffracted by the object will reach the objective

dark-field illumination DENDY (1960) proved that in the range of visually countable grain densities there is a linear relationship between the number of grains and the luminous flux they diffract into the objective.

However, transmitted-light dark-field photometry has a number of fundamental drawbacks. Without coarse displacement of the substage dark-field condenser it is impossible to view the autoradiographs in a bright field. This considerably complicates evaluation by certain morphological criteria. Incidentally, the luminous flux diffracted into the objective by the silver grains is greatly influenced by the condenser setting. Even a minute displacement requires careful readjustment. Allowance has to be made for the light absorbed by the actual object. We thus have the same limitations for this photometric technique as were mentioned for transmitted-light bright-field photometry. While the method of substage dark-field illumination is suitable for the qualitative assessment of grain densities in autoradiographs (ENGELS, 1965), it has been abandoned as a quantitative technique. The apparatus permitting simultaneous scanning, described by TOLLES (1959), has apparently not been used in autoradiography.

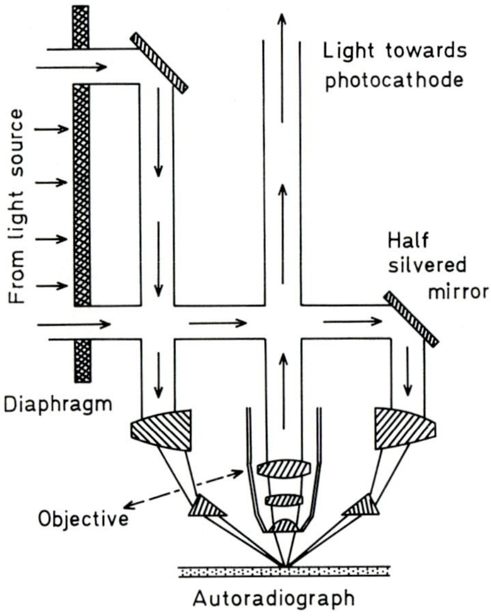

Fig. 1.3. Optical diagram for incident dark-field illumination (Ultropak-type illuminator). A ring condenser is permanently mounted around the objective and illuminates the object at such an angle that only diffracted or reflected light can reach the objective

1.2.3. Photometry by Incident Dark-Field Illumination

With incident dark-field illumination also the light can only reach the objective if its path is modified by the specimen by diffraction and reflection (Fig. 1.3). The light is incident on the autoradiograph from above and can therefore be reflected back into the objective without having touched the radioactive sample. Change-over from incident dark-field illumination to substage bright-field illumination can easily be accomplished without shifting any optical component. This allows the autoradiograph under study to be viewed microscopically whenever desired. As the objective is focused on the object plane to be measured, the illuminating system surrounding the objective is automatically positioned at a constant distance from the object. Photometry by incident dark-field illumination is superior to photometry by vertical bright-field illumination if large areas of silver grains have to be measured at relatively low magnification. At high powers dark-field reflected-light photometry is, however, at a disadvantage because the illuminating beam cannot be stopped down. As a result, background readings are higher than with vertical bright-field illumination (see 1.6.1.3).

Gullberg (1957) described a microscope photometer for measuring silver grains by incident dark-field illumination. This photometer was connected to a scanning system. Since in this method too the luminous flux recorded with low grain densities is proportional to the number of grains, scanning obviously does not hold any advantage over integrating measurement.

Rogers (1964, 1967) has given a detailed description of the pros and cons of the reflected-light dark-field system. He used the Leitz Ultropak system which is designed for viewing specimens without a cover glass. It is known that cover glasses diffract a considerable portion of the incident light, especially with dry systems of high numerical aperture, before it ever reaches the emulsion. Low-aperture objectives and high-aperture immersion objectives are suitable for reflected-light photometry. For single-cell autoradiography, Rogers gives preference to vertical bright-field illumination over incident dark-field illumination.

1.2.4. Photometry by Vertical Bright-Field Illumination

The object is illuminated by the same objective that is used for viewing (Fig. 1.4). In other words, the illuminating aperture is identical to the numerical aperture of the objective. A vertical illuminator is used which reflects the horizontal illuminating beam into the objective by means of a 45° plane parallel plate located above the objective, but does not affect the viewing beam emerging from the objective.

Light is reflected by silver grains many times more intensely than by the stained object structures and the other parts of the object. The result is an image of brilliant silver grains standing out against a dark background. In the range of visually countable grain densities the reflected luminous flux is proportional to the number of grains (ROGERS, 1961, 1964; DÖRMER et al., 1966). For general viewing of the autoradiograph the illumination is simply switched to bright-field transmitted light. The most important advantage compared with incident dark-field illumination is the possibility of stopping the illuminating beam down to the size of the test area. Photometry by vertical bright-field illumination has been used for the evaluation of autoradiographs in which a separation of the nuclear emulsion from the labeled cells or tissues did not seem appropriate (DÖRMER and BRINKMANN, 1970; DÖRMER and MÖLLER, 1968; DÖRMER, REICHART and HUHN, 1968; LÄUCHLI and LÜTTGE, 1968; SCHREIBER et al., 1970).

MERTZ (1969) used vertical bright-field illumination for scanning the bright silver grains with a television camera (QTM, Quantitative Television Microscope). This electronic system allows the measurement of grain diameters. The disadvantages of this setup for routine measurement of grain density, as discussed by MERTZ, are largely identical to those of the system used by DUDLEY and PELC (1953).

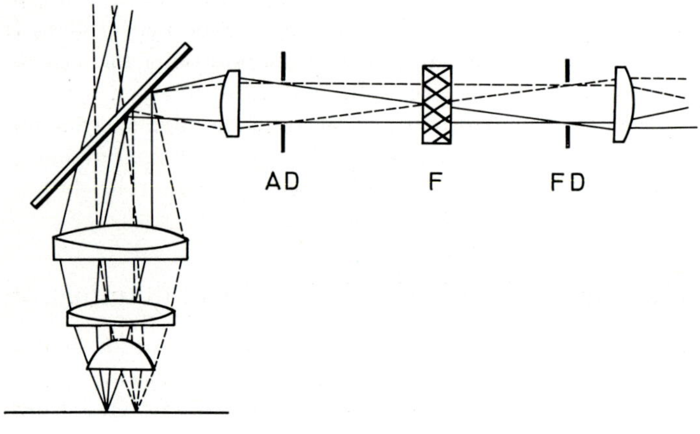

Fig. 1.4. Optical diagram for vertical bright-field illumination. The specimen is illuminated through the objective. AD = aperture diaphragm, F = interference filter, FD = field stop

1.3. *Problems of Proportionality in Quantitative Evaluation*

1.3.1. Relationship between Radioactivity and Grain Density

The autoradiography of biological material is mainly one of beta decay. The following considerations therefore refer to the autoradiography of beta emitters. Grain density in nuclear emulsions has generally been determined with the aid of densitometry, since both theoretically (ARENS, EGGERT and HEISENBERG, 1931; HERZ, 1969; NUTTING, 1913; PELC, 1957) and experimentally (ALTMAN, 1963a; ANDRESEN et al., 1953; ARENS, EGGERT and HEISENBERG, 1931; FRIESER, HEIMANN and RANZ, 1962; JAMES and FORTMILLER, 1961) there is a linear relationship between grain density and the recorded blackening. While there are deviations from this linearity in the case of greatly varying grain sizes in a photographic emulsion (KLEIN, 1958), it is precisely the nuclear emulsions used in the field of autoradiography which distinguish themselves by homogenous grain size (HERZ, 1969), so that we may assume at least an approximately linear relationship.

Numerous experiments have proved that in the range of low blackening there is a linear relationship between the number of beta particles and the resulting blackening (ANDRESEN et al., 1953; BOTHE, 1922; DUDLEY, 1954; GLOCKER, 1960; PODDAR, 1955; TAYLOR, 1956). With higher grain densities, however, the exponential character of this relationship becomes apparent. The deviation from linearity can be explained by the coincidence of beta particles (PERRY, 1964). By this we understand the absorption of the energy of more than one beta particle by one silver halide crystal. The probability that more than one particle transmits its energy to one and the same silver halide crystal increases with the number of crystals hit. PERRY (1964) expressed this relationship in the form of an exponential function:

$$n_{obs} = n_{max}(1 - \exp\{-n_{true}/n_{max}\}),$$

where

n_{obs} is the number of grains observed per unit of area,

n_{max} the maximum number of grains that can be blackened per unit of area,

n_{true} the number of grains per unit of area to be expected in the absence of coincidence.

From this equation it can be deduced that the difference between n_{obs} and n_{true} becomes all the smaller, the smaller n_{obs} is in relation to n_{max}. Hence an approximately linear relationship to the radiation exposure for low grain densities follows. For the grain densities involved in microautoradiography, in which a blackening of 0.3 is not generally exceeded, a linear correlation may be assumed between the number of beta decays and the number of silver halide crystals that have become developable. However, if single layers of grains are used as a nuclear emulsion, the exponential nature of the relationship becomes obvious already at very low grain densities (SVEDBERG and ANDERSSON, 1921). In macroautoradiography also, the blackening involved is such that a linear relationship to the radiation exposure cannot be taken for granted. Special conditions are encountered in tritium autoradiography where n_{max} is only a small portion of the silver halide crystals per unit of area, due to the very short range of the beta particles. Accordingly, the coincidence of beta particles has been proved to exert a noticeable influence in tritium autoradiography (DÖRMER, 1967b; DÖRMER and BRINKMANN, 1968).

1.3.2. Relationship between Grain Density and Photometric Response

In order to compare the measuring techniques with the different types of illumination, AR-10 stripping film was mounted on specimen slides, dried and exposed to x-ray radiation at 60 kV. It was then developed for

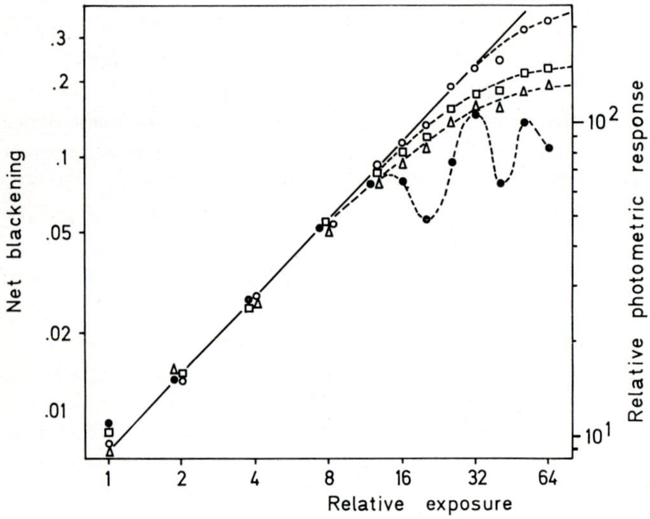

Fig. 1.5. X-ray exposure of AR-10 stripping-film preparations, plotted against photometer readings of grain densities with different types of illumination. Symbols: ○---○ substage brightfield illumination (measurement of blackening); □---□ vertical brightfield illumination; △---△ incident dark-field illumination; ●---● substage dark-field illumination

six minutes in Amidol at 18 °C. The blackening produced by x-rays is comparable to that caused by beta emitters (BECKER, 1962; HERZ, 1969). The results were plotted to a logarithmic scale on both axes (Fig. 1.5) to allow comparison.

1.3.2.1. Response in Substage Bright-Field Illumination

The curve obtained by densitometry shows the longest linear section. The densitometric measurements were performed both in a double-beam microdensitometer (JOYCE-LOEBEL, Mark III) and in a macrodensitometer (BALDWIN Densitometer) and produced identical results.

The extension of the range of linearity between blackening and radiation exposure depends, on one hand, on the type of nuclear emulsion used (COBB and SOLOMON, 1948; ODEBLAD, 1952). On the other hand, the length of the linear section can be strongly influenced by processing conditions (COBB and SOLOMON, 1948; JAMES and FORTMILLER, 1961; ODEBLAD, 1952). It is quite possible that the deviation of the densitometric curve from linearity, evident in Fig. 1.5, could have been prevented in the range measured by selecting a different developer.

1.3.2.2. Response in Substage Dark-Field Illumination

After a relatively short linear section of the curve there is a disproportionality which would make further measurement useless. The same phenomenon of disproportionality was encountered in measurements by vertical bright-field illumination (DÖRMER, 1967a), though at considerably higher grain densities. The reason for this response is unknown. In a second experiment under identical conditions an attempt was made to reproduce the disproportional section. However, this proved impossible; instead, a saturation curve was found which very soon became horizontal. This amply demonstrates that photometric results should be thoroughly tested for reproducibility and linearity. Darkroom conditions must by all means be standardized.

1.3.2.3. Response in Incident Dark-Field Illumination

Here also we first have a linear relationship which, however, soon takes the form of a saturation curve. The transition occurs later than in transmitted-light dark-field photometry, but earlier than in reflected-light bright-field photometry.

1.3.2.4. Response in Vertical Bright-Field Illumination

After densitometry, this is the photometric method which yields the most favorable relationship to radiation exposure. Loglog plotting, however, suggests a rigorous linearity in the lower part of the curve that does not exist in reality. This is borne out by Fig. 1.6: After photometric measurement of the AR-10 stripping preparations for Fig. 1.5, the grains were separated from the emulsion by boiling, diluted with gelatine free from silver grains and smeared on slides as a new emulsion. In these specimens the number of grains per area were counted visually and the thickness of the emulsion in the count areas determined interferometrically. The dilution factor being known, this allowed the grain density in the original stripping-film preparations to be determined. This was then plotted against the grain density resulting, after conversion, from the photometric results obtained by vertical bright-field illumination. By approximation, an exponential function of the following form can be derived from these results:

$$n_{phot} = n_{max}(1 - \exp\{-n_{true}/n_{max}\}),$$

where

n_{phot} is the grain density computed from the photometric results,

n_{max} the grain density computed from the maximum photometric results (photometry in the saturation range), and

n_{true} the actual grain density.

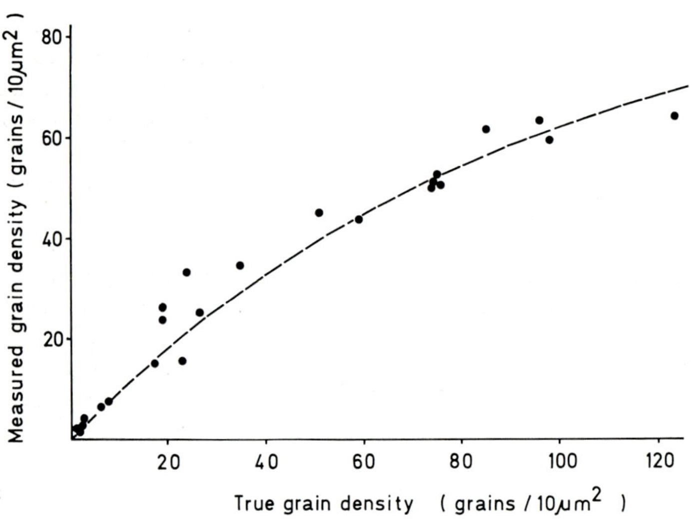

Fig. 1.6. True grain density plotted against the grain densities computed from the reflected-light bright-field measurements. Assuming an exponential function, a curve for an n_{max} of 95 silver grains per 10 μm^2 was plotted (dashed line)

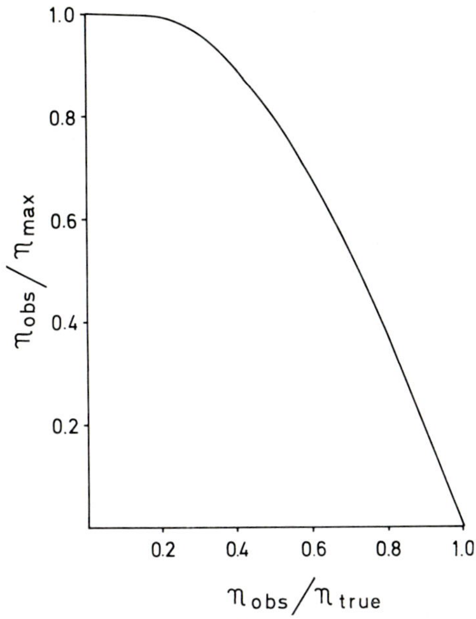

Fig. 1.7. Observed grain density as a fraction of true or expected grain density, plotted against the observed grain density as a fraction of maximum measurable or inducible grain density (after PERRY, 1964). For low grain densities the curve is approximately linear: $y = z(1 - x)$; this linear section may be used to convert exponential into linear relationships

The curve in Fig. 1.6 was constructed for an n_{max} of 95 silver grains per 10 μm². Previous measurements (DÖRMER, 1967b) had revealed an n_{max} of approx. 115 grains per 10 μm². Considering the aforementioned dependence of photometric results on photographic conditions, this difference falls within the range of deviations that was to be expected. In view of the results shown in Fig. 1.6 the linearity with radiation exposure up to 70 grains per 10 μm² (DÖRMER, 1967b) must be considered as apparent only. Due to the influence of darkroom processing it cannot be taken as a constant of the AR-10 stripping film.

It may be deduced from the results shown in Fig. 1.6 that in reflected-light bright-field photometry as in all other forms of silver-grain photometry (with the exception of densitometry) there is no range of true linearity between the grain density actually present in the emulsion and the one computed on the basis of the measurement. However, the exponential function exhibits so little curvature in its first section that in this range of grain densities we may neglect the deviation from linearity. On the other hand, if measurements are made at higher grain densities and if optimum accuracy is required, a method may be used which results from a simplification of the exponential function.

PERRY (1964) has shown that the exponential function developed by him yields a relationship between n_{obs}/n_{max} and n_{obs}/n_{true} that is approximately linear for large values of n_{obs}/n_{true}, i.e. for low grain densities (Fig. 1.7). This form of representation had been chosen in order to calculate the coefficient of coincidence in tritium autoradiography (DÖRMER, 1967b). The approximately linear section in Fig. 1.7 can be described by the function $Y = Z\,(1 - x)$. A Z-value determined in this manner for the series of points in Fig. 1.6 lies around 180. The actual grain density n_{true} can then be easily computed from the measured n_{phot} using the formula:

$$n_{true} = \frac{Z \times n_{phot}}{Z - n_{phot}}$$

A detailed derivation of this equation is given in DÖRMER (1967b).

1.3.3. Comparison of Different Types of Illumination

It follows from the relationship between the photometric measurement of grain densities and the true grain density that by far the most favorable photometric method is densitometry. The linearity between densitometric measurement and radiation exposure of the film is disturbed only by the phenomenon of coincidence of radioactive particles. In appropriate photochemical conditions (HERZ, 1969), it is possible to create a linear relationship for a range that is large enough at least for microautoradiography. The limited usefulness of densitometry in autoradiography is due to the fact that it is impossible to measure stained or even unstained tissue together with the superimposed silver grains. In all those cases in which the two components of the autoradiograph can be separated, densitometry is the method of choice.

It may be simpler to measure the absorption of light by substage brightfield illumination instead of blackening, but the absorption of the light by the silver grains is related exponentially to the radiation exposure. This is also obvious from the results obtained by FISCHER and WERNER (1968). It therefore only appears useful in the case of low grain densities to measure absorption instead of blackening.

Reflected-light dark-field photometry can be used to advantage if on the one hand the nuclear emulsion cannot be separated from the labeled specimen and if, on the other, large areas have to be measured. It should be taken into account, however, that the range of linearity between radioactive exposure and photometric response is notably smaller than in reflected-light bright-field photometry.

The latter method is best suited for single-cell autoradiography. According to the results available to date (see also Fig. 1.22), the error caused by non-linearity should be considerably smaller, up to a density of about 30 grains per 10 μm^2, than the error introduced by other factors (see 1.7.1). A correction of photometric results will thus hardly be necessary up to this grain density. If the grain density does not vary excessively but is required to be above this limit of 30 grains per 10 μm^2, the error due to non-linearity can be eliminated by the simultaneous exposure of radioactive standard sources of a specific activity comparable to that of the specimen.

A great deal has been published on the production and use of such radioactive standards, to which only general reference can be made here: ANDRESEN et al., 1953; AXELROD and HAMILTON, 1947; BERLIN and ULLBERG, 1963; BLACKETT, KEMBER and LAMERTON, 1959; DOMINGUES, SARKO and BALDWIN, 1956; DÖRMER and BRINKMANN, 1970; DUDLEY and DOBYNS, 1949; JOWSEY, LAFFERTY and RABINOWITZ, 1965; KUTZIM, 1963; MAMUL, 1956; RITZEN, 1967a, 1967b; VAUGHAN and OWEN, 1959; WASER and LÜTHI, 1962. The simultaneous exposure of radioactive standards in the autoradiographs is indispensable for determining the absolute radioactivity in cells or tissue.

There is no sphere in which the use of substage dark-field illumination would present any advantage over another method. In view of the drawbacks mentioned above, its use in quantitative autoradiography can be dispensed with.

1.4. *Design of a Microscope Photometer for Reflected-Light Bright-Field Work*

Since the photometric measurement of autoradiographs by vertical bright-field illumination has proved to be of advantage in several respects, a more detailed description of this method will be given in the following. Only a small portion of the methodological prerequisites is limited exclusively to photometry by vertical bright-field illumination; the major part applies by analogy to the use of other illuminating techniques. Since the author primarily used a Leitz photometer for his work (WEBER, 1965; DÖRMER, 1967a), the description refers above all to that instrument (Fig. 1.8). In the meantime, however, comparable microscope photometers have also been brought out by other optical companies.

1.4.1. Optical Components

The light source used for vertical illumination is a mercury vapor lamp located on one side. The power supply of the lamp must be properly stabilized. The illuminating beam complies rigorously with the KÖHLER method (see Fig. 1.4). An aperture diaphragm is not required because it is advisable always to use the maximum aperture for illumination (see Fig. 1.12). In

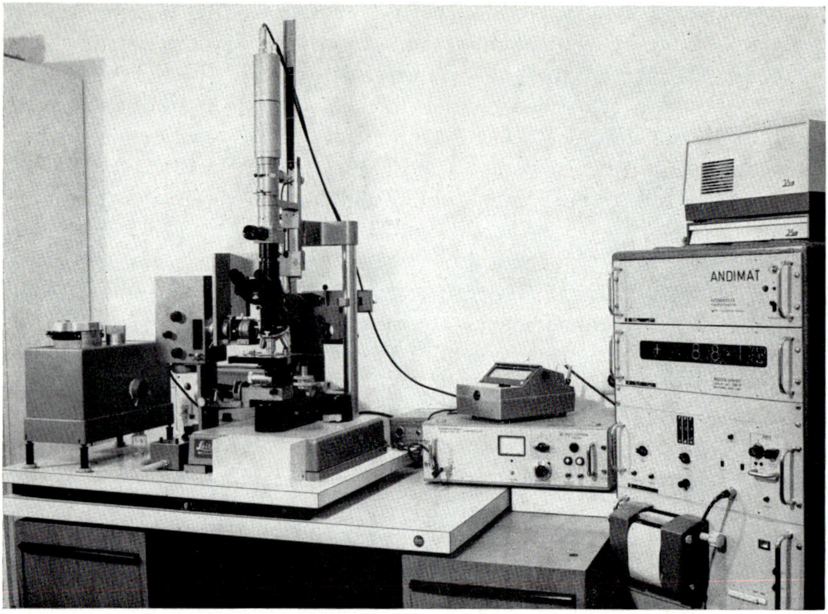

Fig. 1.8. General view of a single-beam photometer for reflected-light bright-field measurement with analog-to-digital converter, digital display, punching and printing system

order to eliminate unwanted flare in the objective, a stray-light stop is provided in the back focal plane of the objective. An interference filter can be moved into the vertical illuminating beam. The vertical illuminator contains a 45° plane-parallel plate reflecting the light into the objective. A quintuple revolving nosepiece allows observation at different magnifications.

The light reflected by the silver grains either reaches the observer's eye through a binocular body or is projected into the plane of the measuring aperture by means of an interchangeable projection eyepiece (Fig. 1.9). The measuring aperture is illuminated from above by a low-voltage lamp. The image of the illuminated measuring aperture can be viewed through a telescope together with the image of the reflecting silver grains (Fig. 1.10). The brightness of measuring-aperture illumination can be controlled by means of a transformer so that the autoradiographic detail stands out clearly

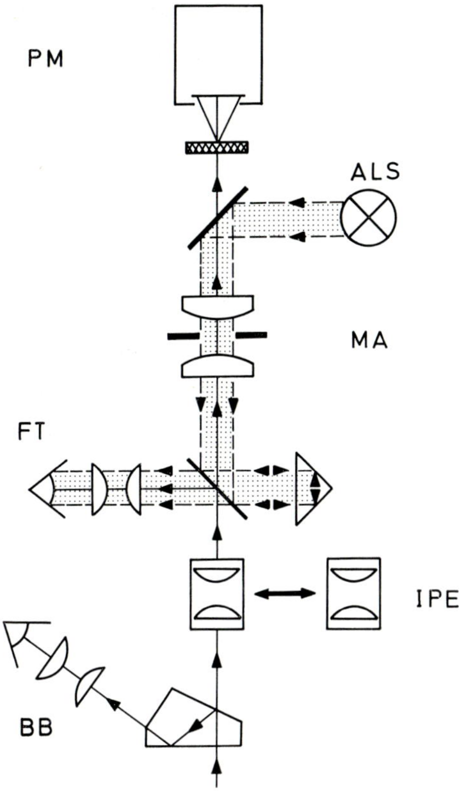

Fig. 1.9. Light path in photometer section with possibility of masking down the measuring beam while viewing the entire microscopic image. PM = photomultiplier, ALS = auxiliary light source, MA = measuring aperture, FT = focusing telescope (for checking the test area) IPE = interchangeable projection eyepiece, BB = binocular body for general viewing

from the illuminated field. While they are viewed through the telescope, the object details of interest can be masked down with the aid of interchangeable iris or rectangular stops. An interference filter can be inserted below the photocathode.

A standard low-voltage lamp with substage condenser serves for ordinary viewing by substage bright-field illumination. When selecting a condenser, it should be remembered that some types may reflect the light back into the objective. Such a condenser has to be removed from the light path before every measurement (Rogers, 1961). Other types of condensers do not have this property and may be left in position during measurement, provided that they are properly centered (Dörmer et al., 1966). The transmitted light required for general viewing can be interrupted for measurement or cut in for observation by means of a swing-in mirror or photographic shutter in the base of the microscope.

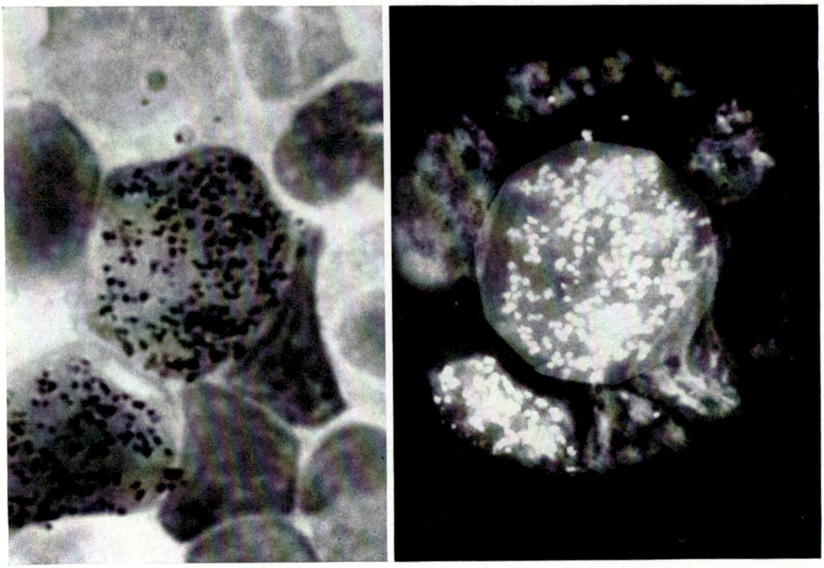

Fig. 1.10. Microscopic image seen through telescope a) with general substage bright-field illumination, b) vertical bright-field measuring illumination. The central, homogeneously illuminated area is equivalent to the size of the measuring aperture in the object plane. The entire illuminated area is determined by the size of the field stop which was here opened wider than would be suitable for measurement. The reflectance of the unlabeled cells can be clearly recognized (micrograph taken without an interference filter in the illuminating beam)

1.4.2. Electronic Equipment

If in an autoradiograph the reflectivity of individual silver grains is assumed to be constant in a statistical sense, the reflected luminous flux is a function of the grain number and the luminous intensity used for illumination. If the illuminating beam and the incident luminous flux are kept constant, then the luminous flux reflected can be considered a relative measure of grain density. However, it is not the reflected flux that is displayed, but the luminous flux converted into a measurable current by means of a photomultiplier. Therefore, if photometer units are to serve as a relative measure of numbers of silver grains, the electronic amplification by the photomultiplier must be kept constant in addition to the illuminating beam and the light intensity. (Donaldson, 1958; Herrmann, 1965) It is therefore necessary to use highly stabilized power supply units for the mercury vapor lamp and for generating the high voltage required for the photomultiplier. In addition, conditions in the author's laboratory made it indispensable to insert a magnetic voltage stabilizer with an input of 1,000 VA before all stabilized instruments (see also PREISSER, 1965).

1.4.3. Recording Equipment

The light converted into a measurable current can be recorded by means of a galvanometer, a pen recorder or an analog-to-digital converter with digital readout. The equipment used in the author's laboratory is an analog-to-digital-converter with digital readout in conjunction with a printing and tape-punching system. The data stored on punched tape are then processed and plotted with the aid of a computer. In order to allow various kinds of data on a tape to be processed separately in the computer, a switch is provided which permits the measurement to be identified by eight different class characters.

1.5. *Preparing the Autoradiographs*

It should be made a rule to treat all autoradiographs of a series, from the application of the film till the end of the staining, identically and, if possible, simultaneously. This hard and fast rule may be neglected only if reference densities are provided on the slides. But even then identical treatment of autoradiographs will only be of advantage.

1.5.1. Selecting a Nuclear Emulsion

Unless tritium is used as an isotope, quantitative autoradiography calls for nuclear emulsions of constant thickness. In the case of stripping film, this condition is met satisfactorily. Stripping films for the most varied nuclear emulsions are commercially available. A listing is given by BASERGA (1967) and HERZ (1969). Emulsion thickness of 5 μm and less are well suited for reflected-light photometry. In the case of excessive emulsion thickness, a more pronounced increase of the focusing error must be expected (see 1.6.4). After development and the elimination of unexposed silver bromide, a 5 μm emulsion is still approx. 2.5 μm thick.

For reflected-light photometry, the mean grain size should not be chosen too small. When silver grains with a mean diameter of 0.07 μm in a Scientia NUC-3.07 emulsion (Agfa-Gevaert) are illuminated by reflected light, they exhibit only very weak reflection, while their absorption in transmitted light is considerably higher. Nuclear emulsions with grain diameters between 0.15 and 0.2 μm (Ilford K-2, Agfa-Gevaert NUC-7.15, Kodak AR-10) and emulsions with larger grain diameters are, however, well suited for reflected-light photometry.

1.5.2. Exposure

It has been found of advantage for photometric purposes to use higher grain densities than would normally be advisable for visual counting (DENDY, 1960; DÖRMER et al., 1966). On the other hand, it is not possible to choose excessive grain densities because this introduces the problem of non-linear photometric response (see 1.3) and because excessive grain density will complicate morphological assessment of the underlying structures. Densities of between 5 and 25 grains per 10 μm^2 have proved to be most appropriate in the author's investigations.

1.5.3. Development

The size of the silver grains in an autoradiograph can be influenced by the type of developer, the development time and temperature. Since the reflected luminous flux is a function of grain size, all processing operations should be kept strictly constant. Thorough removal of residual developer and fixing bath from the emulsion by sufficient rinsing for about two hours is particularly important.

Any crystals left in the emulsion will reflect very strongly and may result in an increased and partly very irregularly distributed noise component. If the emulsion allows it, the rinsing process should be followed by brief washing in destilled water.

If radioactive standard sources are exposed together with the autoradiographs, special attention should be paid to development. The majority of previously used standards consisted of substances such as methacrylate, which are relatively foreign to tissue. Depending on the type of developer used, smaller silver grains may be formed above such standards than above cells and tissue.

MERTZ (1969) and ROGERS (1967) have proved that the silver grains formed above the background are smaller than those above the cells. Both authors used hydroquinone developer (D-19, for composition see BASERGA and MALAMUD, 1969). The author's own experiments with D-19 developer revealed, for 100 silver grains above a ^{14}C-methacrylate standard, a photometer reading of 103.5 ± 1.0 and above ^{14}C-labeled thymus cells a photometer reading of 122.4 ± 1.5 ($p < 0.01$). However, if identical autoradiographs were processed in Amidol developer, 100 grains above the standard source gave a reading of 72.6 ± 0.6 and above the cells one of 72.9 ± 0.8 photometer units.

It is obvious that Amidol developer (D-170; for composition see BASERGA and MALAMUD, 1969) was well suited for determining the absolute radioactivity in the cells of these autoradiographs, while hydroquinone developer was unsuitable. A photometric comparison of grain densities above very different radiation sources should not be made without an occasional check on the mean grain size (for procedure, see 1.7.2).

1.5.4. Staining

All dyes have the property of reflecting incident light. This property has been studied in detail by WESTPHAL (1963) for a large number of dyes. Above all, those wavelengths are reflected which are complementary to the intrinsic color of the dye. While it is true that the reflectivity of individual dyes varies greatly, it may be said in general that the dye reflectance in an autoradiograph increases with the dye concentration (DÖRMER et al., 1966).

Since the dye reflectance is largely responsible for background brightness in reflected-light bright-field photometry, all provisions designed to reduce the stain intensity in the nuclear emulsion will help to reduce the background brightness. Conditions are particularly favorable if the specimens can be stained before exposure of the nuclear emulsion, for example by FEULGEN stain. It should be remembered, however, that the risk of chemography makes only few dyes suitable for this procedure (for details see BASERGA and MALAMUD, 1969; FISCHER and WERNER, 1971; cf. also 5.11).

If staining before exposure of the autoradiographs is not possible, the dye often can be largely removed from the emulsion after the autoradiograph has been stained. With histological specimens, for example, the following procedure with MAYER's hemalum, without counterstaining, has given good results: the autoradiographs are stained in the freshly filtered solution for 15 min and then briefly rinsed in tap water and dried. Next follows differentiation in HCl alcohol (96 parts 70% alcohol + 4 parts 1 nHCl) for 3 to 8 min, rinsing in tap water and air drying. Giemsa stain, in which the normally pronounced reflectance of the Eosin component is largely eliminated, can be obtained by the following procedure: 40 min staining of the autoradiographs in a filtered solution of 5 parts of Giemsa stock solution (Messrs. Merck, Darmstadt) and 95 parts phosphate buffer, pH 6.4. This is followed by 24 hrs drying, whereupon the emulsion is decolorized in a solution of 30 parts absolute alcohol and 70 parts phosphate buffer pH 6.4. at 15° C. Without a final rinse, the autoradiographs are then air-dried.

Secondary decolorization of the emulsion has another reason: the light striking the silver grains and reflected by them into the objective is absorbed by the dye in the emulsion. The degree of absorption is determined by the dye concentration, the wavelength used for measurement, the absorption characteristics of the dye at that wavelength and finally the thickness of the emulsion. On account of the last factor, the staining of dipping emulsions is not advisable because the thickness of the emulsion is not generally sufficiently uniform.

The type of dye and the wavelength to be used for measurement should preferably be selected so that the dye has minimum absorption at the wavelength used. This will automatically insure minimum reflectance (see 1.6.3).

If autoradiographs stained with different dyes or autoradiographs stained with different intensity are to be studied together, an estimation of light absorption is indispensable. This can be done by measuring the absorption of light in substage bright-field illumination in areas of the autoradiograph in which there is no object. There is a constant ratio between transmitted-light bright-field absorption and reflected-light bright-field absorption, which is about 1 : 1.67 for Kodak AR-10 (Fig. 1.11). If necessary, a correction for light absorption should be applied in the various autoradiographs. Such a correction is not required if the autoradiographs include standard exposures and if the emulsion thickness is relatively constant.

1.5.5. Mounting

When stripping film is used, the specimens are left uncovered. Due to its roughly 10 μm thick gelatine layer above the silver grains, this type of film prevents mechanical damage to the photographic emulsion proper. ROGERS (1961) has shown that the noise reading is noticeably increased by the cover glass and the mounting medium. This agrees with the author's own experience. In addition, the noise in an autoradiograph is not constant as long as there is the least residual moisture in the mounting medium. With large cover glasses, it may be many days before all the medium is completely dry.

Care should be taken to insure that the underside of the specimen slide is completely free from immersion oil during photometric measurement. Even an extremely thin oil film on the underside may act as an optical boundary face and increase the noise reading.

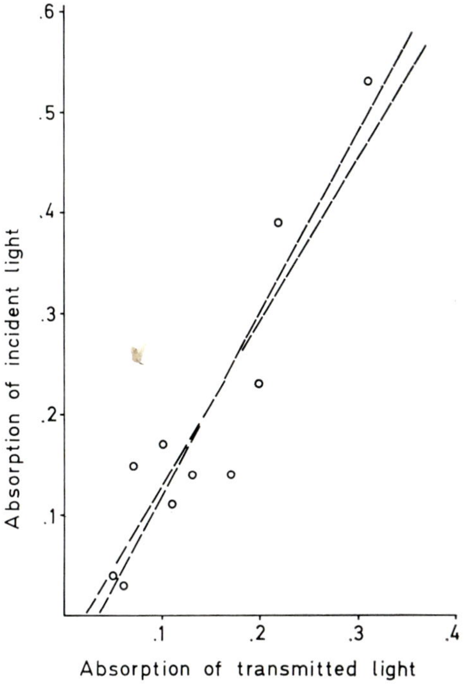

Fig. 1.11. Light absorption with vertical bright-field illumination (incident light) as compared to substage bright-field illumination (transmitted light) in AR-10 stripping film stained with MAYER's hemalum

1.6. *Procedure for Reflected-Light Bright-Field Photometry*

1.6.1. Aperture Settings

1.6.1.1. Aperture Diaphragm

The autoradiographs are illuminated with wide-open aperture diaphragm. This gives the most favorable signal-to-noise ratio (Fig. 1.12). We here have the opposite case to that of transmitted-light bright-field photometry where the smallest possible aperture is used for illumination. When the aperture diaphragm in the vertical illuminator is fully open, care should be taken to prevent the illuminating beam from striking the edges of the lens elements in the objective. Should this happen, it is advisable either to reduce the illuminating aperture accordingly or to make a stray-light stop of black paper and mount it near the back focal plane of the objective to reduce illumination of the lens edges. In the photometer used by the author, this simple aid made it possible to reduce the background brightness by almost 50%.

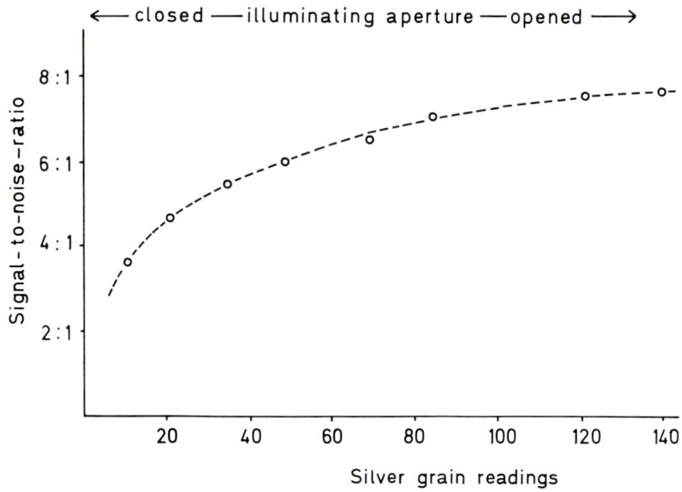

Fig. 1.12. Effect of aperture diaphragm on signal-to-noise ratio

1.6.1.2. Measuring Aperture

Measurements with apertures of fixed size are always simpler and quicker than measurements with apertures of variable size. However, with numerous biological objects it is desirable to adapt the size of the measuring aperture to the structures concerned — for example, when grain densities have to be measured above cells of greatly varying size. In this case, it is a great advantage if the object can be viewed even within the test area. However, this requires the insertion of a complex optical system as shown in Fig. 1.9 (Weber, 1965).

Suppression of the light reflected by silver grains outside the test area can be achieved only in part. This becomes clear when the intensity of the light is measured above and outside an isolated grain (Fig. 1.13). The light contributed by the silver grains outside the test area is partly due to scattering by optical components. It follows from this that as few optical components as possible should be located in the light path. On the other hand, however, this additional light from outside the test area is due to the fact that in vertical bright-field illumination every silver grain produces diffraction patterns. Such diffraction patterns from grains outside the test area can always find their way into the test area, in spite of the measuring aperture. The size of the diffraction patterns increases as the grains are defocused. Consequently, a higher component of light from outside the test area is to be expected inside that area in the case of defocused grains (Rogers, 1967). The setup used by Dörmer et al. (1966) was equivalent to defocusing be-

Fig. 1.13. Light measured above and outside a silver grain in vertical bright-field illumination. Diameter of measuring aperture approx. 0.7 μm

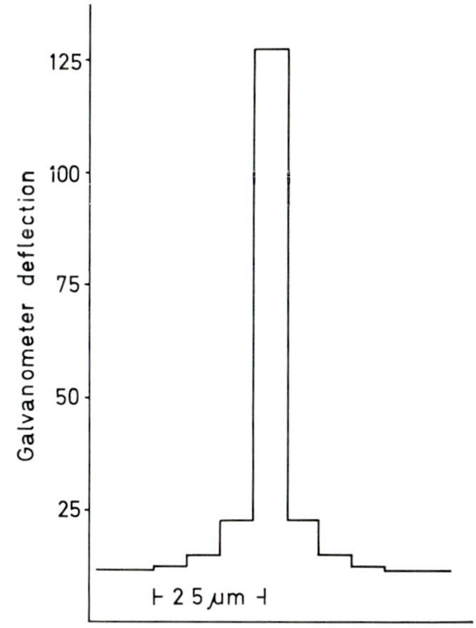

cause the grains were focused in white light and measured in blue light. The stray-light component was therefore considerably larger than in the focused measurement illustrated in Fig. 1.13.

If accumulations of grains are to be measured, as is the case, for example, with ^{3}H-labeled cell nuclei, the measuring aperture should always be chosen slightly larger than the area actually covered by the grains. If the grain density is uniform or approximately uniform and a comparatively small area is measured, it may be assumed that as much light will stray into the area as is lost from it.

The size of the test area will always be selected in accordance with the conditions in the autoradiograph. If the grain density is very low, a larger test area will have to be chosen than for a higher grain density. In no case should the test area be chosen greater than is necessary for the measurement, in order to keep the noise component down as far as possible.

1.6.1.3. Field Stop

The size of the field stop determines the size of the illuminated area in the object plane. The latter, in turn, is decisive for background brightness (Rogers, 1964; Dörmer et al., 1966; Dörmer, 1967a). Fig. 1.14 shows that background brightness increases linearly with the radius of the illu-

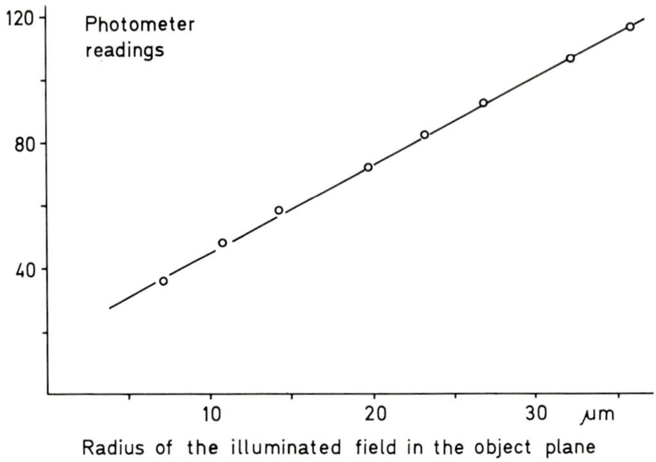

Fig. 1.14. Effect of aperture of field stop on the noise reading in a test area of constant size

minated field if the size of the test area is constant. It follows from this that the illuminated area in the object plane should be kept as small as possible.

The image of the field stop on the object will be all the smaller, the higher the magnification of the objective used. This explains why only high-power objectives are suited for reflected-light bright-field photometry, since only these allow the field stop to be imaged sufficiently small in the object plane to give a satisfactory signal-to-noise ratio. However, the optional use of 6× to 25× projection eye pieces still leaves the possibility of different degrees of secondary magnification.

Due to the thickness of the emulsion, the image of the field stop in the object plane is not in perfectly sharp focus. If a perfectly sharp field-stop image is desired with the high aperture objectives used, there is a risk of choosing an emulsion whose thickness is less than the maximum range of the beta rays of tritium. Due to the marginal blur which the field-stop image exhibits, for example, on AR-10 stripping film, it is necessary to choose the diameter of the field stop in the object plane slightly larger than that of the measuring aperture. If the size of the test area is to be varied in a series of measurements, the size of the field stop should be chosen so that the largest test area of the series is just fully enclosed. The field stop should then be locked in this setting and not disturbed during the entire set of measurements.

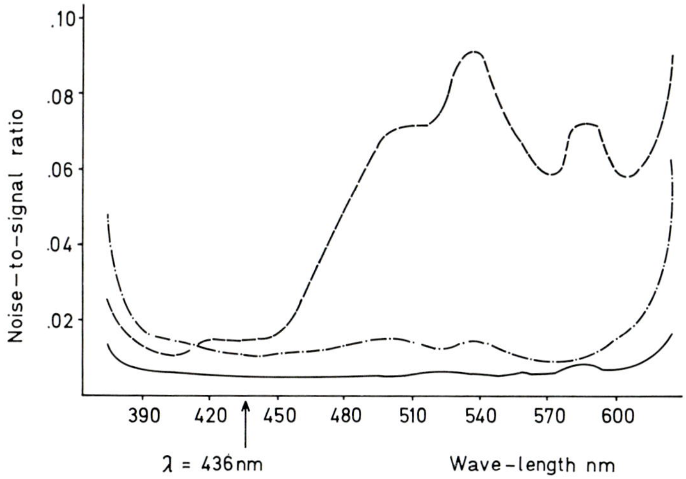

Fig. 1.15. Spectral reflectance of dye expressed as noise-to-signal ratio, as a function of the wavelength used for illumination. Symbols: —— MAYER's hemalum; –·–·–·– GIEMSA stain without a cell in the test area; ------- GIEMSA stain with cell in test area

ROGERS (1967) prefers to mask down the field exclusively with the field stop and to dispense entirely with the measuring aperture. This offers the advantage of a less complex light path in the photometer and thus of a slightly lower noise reading. However, a drawback of this method is that only extremely thin emulsions can be used to obtain a sharp image of the field stop. Moreover, so much stray light is produced in the vertical illuminator that a certain background brightness remains outside the illuminated portion of the object and will be included in the measurement as there is no measuring aperture. Silver grains in the direct vicinity of a masked-down object detail may still be visible as faintly luminous points. As a result, their light will also be measured if there is no measuring aperture. The experience gathered in the author's laboratory seems to indicate that the use of a measuring aperture holds more advantages than may be offset by the more complex light path it requires. While the versatility of the instrument is increased, the overall error certainly is not.

1.6.2. Selecting a Filter

Fig. 1.15 shows the spectral reflectance of two different dyes. One of them exhibits relatively weak and roughly constant reflectance over the entire spectrum. The other exhibits different reflectance for a stained cell or a blank field. Differences of this kind must be expected in differential staining with several dye components of different affinity to the object details. Fig. 1.15 shows that these differences of reflectance can be offset by the use of a suitable filter, in this case one of 436 nm.

If interference filters are used for measurement it is important to focus at the same wavelength that is used for measurement. If the silver grains were focused in unfiltered light but the measurement made in blue light, considerable focus differences would arise unless the vertical illuminator were perfectly corrected for color. As a result, heavily defocused grains would be measured. The grains would then exhibit wide haloes which might even encroach on the test area. On the other hand, however, the measuring aperture will cut off part of the light from grains at the edge of the test area (see 1.6.1.2).

When grains are measured without an interference filter, the peak sensitivity of the photocathode should be in the green region, because the human eye focuses on the wavelength of green light. The blue component of the reflected light is in this case just as defocused as the red component — which corresponds to the chromatic correction of the vertical illuminator. If a photocathode has its peak sensitivity in the blue region of the spectrum, it will receive the defocused wavelengths of blue light relatively more strongly than those of green light. The sources of error are thus the same as outlined above.

1.6.3. Focusing

The emulsions used in quantitative autoradiography are generally too thick to allow simultaneous focusing of the silver grains in the different layers with the objectives of high numerical aperture suited for reflected-light bright-field photometry. Since the luminous flux reflected by defocused silver grains is smaller than that from grains in the focal plane, the contribution by grains from the different planes in a nuclear emulsion will vary when the objective is focused on a certain plane. Unless precisely the same emulsion plane is focused in every case, a focusing error must be expected. Such a focusing error is also known from transmitted-light absorption photometry (Exner, 1960; Sandritter, 1965).

Fig. 1.16 shows the reflected-light bright-field measurements performed on a ^{14}C-labeled cell in a stripping-film autoradiograph. If all grains are far above or far below the focal plane, the focusing error is considerable. In the focal area between the upper and lower emulsion planes, however, the focusing error is low and roughly comparable to the one encountered in transmitted-light absorption photometry (Sandritter, 1965). As early as 1961, Rogers mentioned the relatively small focusing error in reflected-light bright-field photometry.

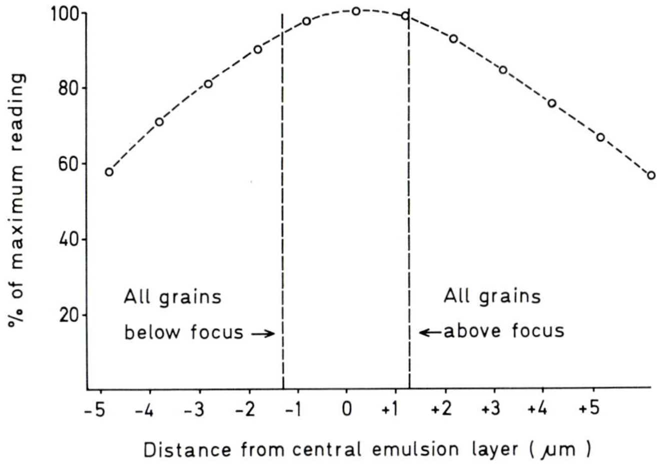

Fig. 1.16. Magnitude of focusing error in reflected-light bright-field photometry of a ^{14}C-labeled cell in AR-10 stripping-film autoradiograph

In order always to focus on the same plane of nuclear emulsion and thus reduce the focusing error to a minimum, we focus on the emulsion plane adjacent to the cells. It is evident from Fig. 1.16 that this procedure involves a focusing error of only a few per cent.

1.6.4. Calibration

As was mentioned in section 1.4.2, it is advisable to insure constant illumination and electronic amplification when measuring silver grains by vertical bright-field illumination. The intervals in which the photometer is readjusted for constant response to the luminous flux incident on the silver grains will have to be chosen in accordance with the properties and quality of the electronic system used. In the past, this adjustment was made with the aid of a labeled cell in a calibration autoradiograph that was moved into the light path at certain intervals. The same cell had to be located every time. Once it had been moved into the center of the test area at a certain setting of the measuring aperture, the high voltage of the photomultiplier was varied until the same galvanometer reading was obtained.

In the meantime, the calibration procedure has been considerably simplified. Beside the objective to be used for measuring the autoradiographs is a turret-mounted calibration objective. This is surrounded by a capsule with a reflector at the bottom (Fig. 1.17). The capsule can be screwed up or

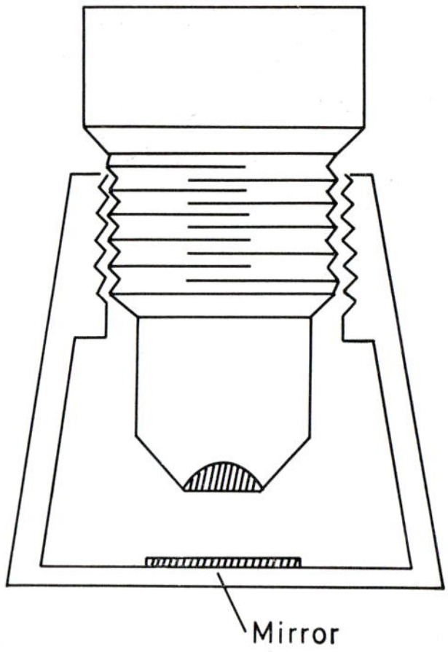

Fig. 1.17. Cross section of a variable reflected-light standard (variable defocusing of a reflector)

down on the objective so that the distance between the reflector and the focal plane of the objective can be varied. This will also vary the intensity of the reflected light. At the beginning of a test series the reflector is first set up at such a distance from the focal plane of the objective that the measurement of the reflected light with a calibration aperture gives about the same readings as the highest grain densities to be measured. The capsule is then fixed in this position by a screw and left undisturbed until the end of the test series. The distance between reflector and focal plane of the objective can be read on the outside of the capsule in relative units (Fig. 1.18).

For repeated calibration of the photometer it is only necessary to move this calibration standard into the light path at certain intervals. Then the photomultiplier high voltage is varied until the meter gives a certain reading for the luminous flux measured with a fixed calibration aperture. This readjustment takes very little time and does not even require the autoradiograph to be shifted on the specimen stage.

In one respect, however, this procedure requires a certain amount of care: I_0 is calibrated with the aid of a defocused reflector, but I is measured in the focused state. Any change in the illuminating beam may affect measurement in and outside the focal plane to a different extent. It is therefore important that the illuminating beam be left perfectly unchanged as long as a test series with such a reflection standard is under way. This refers both to the size and adjustment of the field stop and aperture diaphragm in the vertical illuminator and to the adjustment of the lamp for vertical illumination.

Fig. 1.18. Reflected-light standard with horizontal and vertical calibration in relative units on turret of vertical illuminator

1.6.5. Background Measurement

Even with a well chosen spectrum filter, proper aperture settings and complete absence of silver grains, unlabeled tissue in an autoradiograph will produce a meter reading. This background brightness is composed of stray light introduced by the optical components in the illuminating system, of light reflected from the optical boundary faces between the immersion oil, the emulsion and the slide, and of the light reflected by the dye particles in the stained autoradiograph.

For this reason alone it is necessary to subtract a noise component from the results, which must be determined at identical photometer settings above unlabeled areas, preferably identical unlabeled tissue or cells in the same autoradiograph. In the case of ^{3}H-thymidine labeling, there are always sufficient unlabeled cells between the labeled ones. If labeled RNA or protein precursors are used, it is advisable to transfer both labeled and identical unlabeled tissue to the same slide.

Either a noise reading may be taken for every result, or a regression of noise readings can be established before starting with actual measurements. This calls for a constant size of the field stop and electronic constancy of the photometer. A linear distribution of noise readings will be obtained as a function of the measuring-aperture area. The regression can then be computed for this distribution so that the noise reading for a grain measurement can be calculated from the regression formula as a function of the measuring aperture used or plotted graphically for direct reading. This method has proved to be particularly time-saving when the results are processed in a computer.

If both the field stop and the measuring aperture are kept constant during a test series, the background value of an autoradiograph can frequently be assumed to be a constant. This should first be verified by test measurements at various sites of the autoradiograph. In a favorable case, a mean noise reading with a negligible spread will be obtained, which can be subtracted from all results. In practice, this may be done by presetting a negative meter reading equivalent to the noise reading.

1.6.6. Making Allowance for Geometric Factors

There are border-line cases in quantitative autoradiography, in which the relative concentration of radioactivity can be determined fairly easily, as for instance in tritium autoradiography of sectioned material (Maurer and Primbsch, 1964). With 5 μm, the infinite section thickness is sufficiently assured with tritium. The radiation emerging from a cell and reaching the nuclear emulsion is sufficiently restricted to be quantifiable by counting the grains over the cell. Conditions are also relatively simple if a uniform grain density is distributed over a large area of the specimen. In this case, the grain density found in the center of the area can be considered as representative of the concentration of radioactivity in the source, provided that the specimen thickness is infinite in relation to the range of the radioactive rays. A detailed discussion of these problems has been published by Rogers (1967). Numerous theoretical and experimental investigations have been undertaken to evaluate radioactivity from grain density distributions, even when the diameter of the radiation source to be measured is relatively small in relation to the range of its rays in the film emulsion. (Blackett, Kember

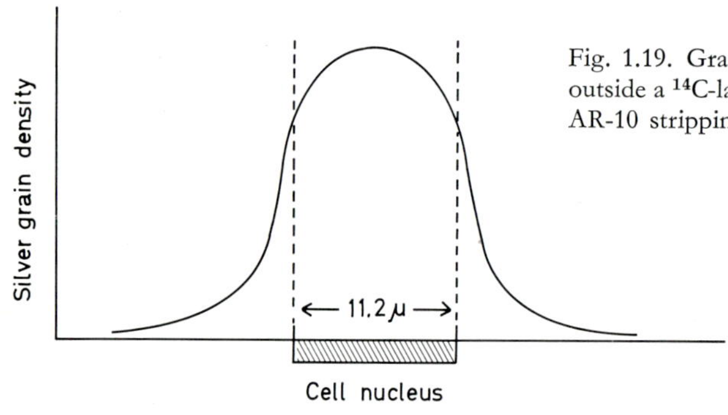

Fig. 1.19. Grain density above and outside a ^{14}C-labeled cell nucleus in AR-10 strippingfilm autoradiograph

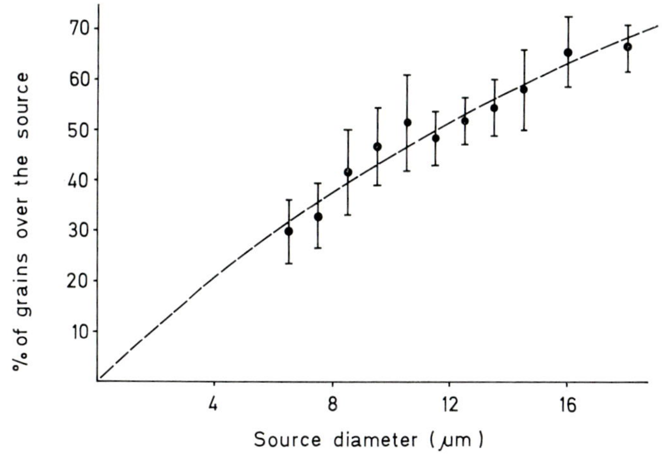

Fig. 1.20. Dependence of percentage of silver grains projected on cylindrical radiation source (^{14}C-labeled cell nuclei in smear) on diameter of radiation sources. An exponential curve was adapted to the results, to express a saturation function

and LAMERTON, 1959; BLEECKEN, 1961; DONIACH and PELC, 1950; GROSS et al., 1951; HERRMANN, HARTMANN and BRUST, 1961; LAMERTON and HARRISS, 1954; MARSHALL, ROWLAND and JOWSEY, 1959; NADLER, 1951, 1953; ODEBLAD, 1952, 1959.)

The problem is illustrated by Fig. 1.19: A ^{14}C-labeled cell nucleus with unlabeled surroundings causes a roughly bell-shaped distribution of grain density in the autoradiograph. If the size of this nucleus were infinite, the number of grains measured outside would be negligible in relation to those above the nucleus. The smaller the nucleus, the larger the percentage of silver grains formed outside the nucleus. The experiment (Fig. 1.20) results in a saturation curve for the percentage of grains above the nucleus with increasing nuclear diameters. It follows from this that a comparable determination of concentrations or amounts of radioactivity is possible only if all the silver grains formed by the source are counted or measured. However, the labeled elements are generally situated so close together in the autoradiographs that their radiation radii overlap.

Attempts were therefore made to compute the total number of grains from the grains overlying the radioactive source. For greater simplicity, the function of Fig. 1.20 was assumed to be linear in the range of nuclear diameters of 6—16 μm (DÖRMER and BRINKMANN, 1970). The equation describing this straight line is $y = 3.43\,X + 9.67$. On checking the magnitude of the error caused by inexact masking-down of the labeled nucleus it is found (Fig. 1.21) that exact masking-down is not critical in determining the total number of grains with the aid of the above equation.

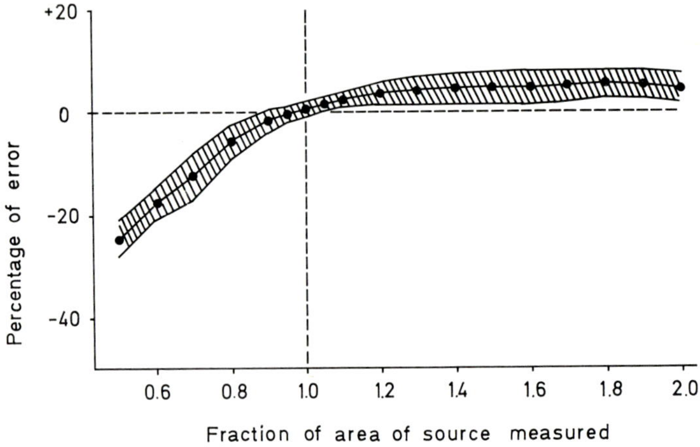

Fig. 1.21. Magnitude of error caused by inaccurate masking-down of cylindrical ^{14}C-radiation sources and subsequent computation of total grain number (mean error $\pm$ standard deviation)

Even when the aperture covers twice the area of the actual nucleus, the total number of grains computed is only 5% too high. It follows that radioactive sources which are neither circular nor angular and which cannot be masked down accurately can also be measured without excessive error. The size of the measuring aperture should preferably be chosen so that it will just enclose the entire labeled source. The area of the aperture will then generally not exceed twice the size of the radioactive source.

If objects with moderately varying diameters of the order of 6 to 16 μm are measured, the measuring aperture may also be fixed in a certain setting (Fig. 1.21) equivalent to roughly the area of the largest of the objects to be measured. A uniform percentage of the total number of grains formed will then be measured in all objects.

It should be mentioned, however, that practical work on the basis of such empirical relationship requires extreme caution and should include frequent checks. It is justified only in fields in which these empirical relationships have been found.

1.7. *Evaluation of Results*

1.7.1. Magnitude of Total Error

The error in the measurement of the silver grains by vertical bright-field illumination is smaller than in visual counting. This was shown by ROGERS (1964) by means of a direct comparison. The standard deviation of a measurement by vertical bright-field illumination found by DÖRMER et al. (1966) was about 3%. The reproducibility of the measurement is even better if the cells are measured with allowance made for their area, as explained in section 1.6.6. Each of three different ^{14}C-labeled cell nuclei were measured 20 times, each measurement being preceded by precise masking-down. The standard deviations for these three cells were 2.1, 2.7 and 0.9%. The degree of reproducibility is thus about identical to that known from absorption photometry.

The total error to be expected in the evaluation of grain densities in autoradiographs is, however, considerably higher and apparently not a function of the magnitude of the measuring error but of the properties of the nuclear emulsion. This will be explained with the aid of the following experiment (Fig. 1.22): 30 uniformly ^{14}C-labeled polymethacrylate sections of different surface activity were mounted on three specimen slides. These specimens were first exposed for 11 days. The grain density in AR-10 stripping film was measured and corrected for different emulsion thickness determined by interferometry. The stripping film was then removed and the procedure repeated with 96 days' exposure. For each of the radioactive specimens the grain densities obtained by the short exposure were plotted against those produced by the long exposure. Since about 50 measurements were made per point, the photometric error may be neglected in this experiment. Fig. 1.22 clearly shows a linear relationship with a certain amount of spread. Considering the combined effect of errors, we can calculate that the standard deviation of this spread is about 10%.

Standard deviations of the same magnitude may be assumed to be present in the results obtained by others who determined blackening or grain density as a function of radioactive exposure (ANDRESEN et al., 1953; AXELROD and HAMILTON, 1947; DÖRMER and BRINKMANN, 1970; KUTZIM, 1963; MARINELLI and HILL, 1948; ODEBLAD, 1952). These standard deviations will probably be due to properties of the nuclear emulsions that were not under complete control in the experiments. From this it follows for the planning and execution of a quantitative autoradiographic study that results should always be based on the investigation of a reasonable number of autoradiographs.

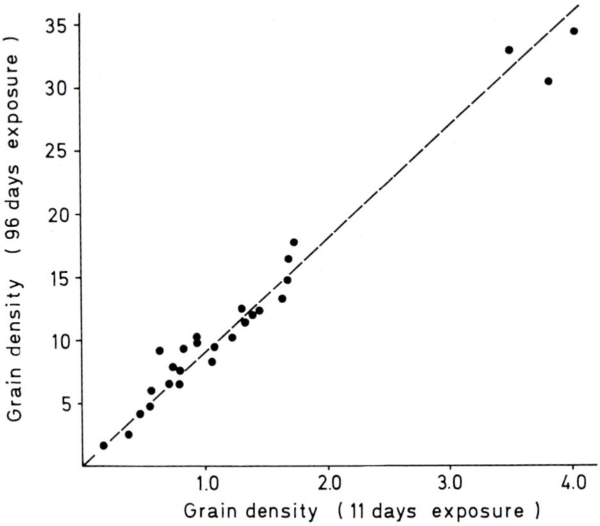

Fig. 1.22. Grain densities above ^{14}C-polymethacrylate sections of different surface activity. The readings of the sections after 11 days' exposure were plotted against those after 96 days' exposure

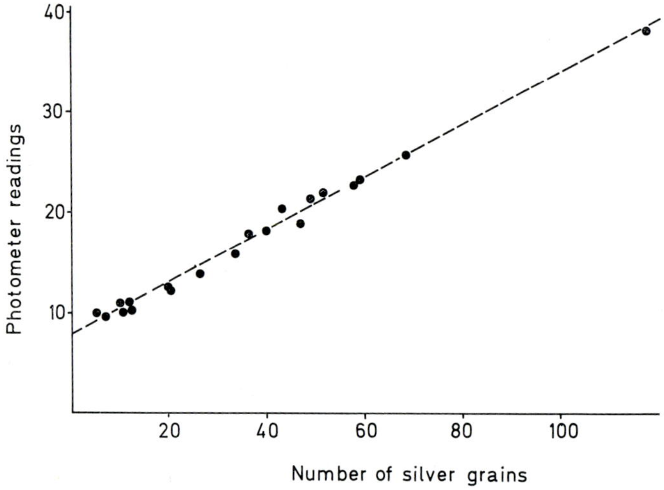

Fig. 1.23. Measurement and visual counting of silver grains for determining the factor for conversion of photometer units into grain numbers

1.7.2. Converting Photometer Readings into Grain Densities

Under certain conditions it may be necessary to convert the photometer readings into grain densities. This procedure is recommended above all if autoradiographs have to be compared with others developed under different conditions, since there is a risk of different grain diameters in the different autoradiographs.

For conversion we have to know the size of the test area in the object plane that was used for the measurements. Using the same photometric calibration as was employed for the measurements, various grain densities are then measured with constant aperture and the grains per test area counted visually. Twenty of these tests are usually sufficient for determining a regression between grain number and photometer reading with satisfactory correlation (Fig. 1.23). Reflected light should preferably be used for the visual counting of silver grains (see WESTPHAL, 1963). Such a straight regression line does not pass through the zero point, since no noise readings were substracted from the measurements. The slope of the straight line determined in Fig. 1.23 is the factor by which the photometer readings have to be divided in order to express them in grain numbers. If the grain numbers are referred to the test area, grain density is obtained.

References

ALTMAN, J.: Regional utilization of leucine-H^3 by normal rat brain: Microdensitometric evaluation of autoradiograms. J. Histochem. Cytochem. **11**, 741—750 (1963a).

ALTMAN, J.: Differences in the utilization of tritiated leucine by single neurones in normal and exercised rats: An autoradiographic investigation with microdensitometry. Nature **199**, 777—780 (1963b).

ANDRESEN, N., CHAPMAN-ANDRESEN, C., HOLTER, H., ROBINSON, C. V.: Quantitative autoradiographic studies on the amoeba chaos chaos with ^{14}C. Compt.-rend. Lab. Carlsberg, Sér. chim. **28**, 499—528 (1953).

ARENS, H., EGGERT, J., HEISENBERG, E.: Zusammenhang zwischen Schwärzung, Silbermenge, Deckkraft, Kornzahl und Korndimension entwickelter photographischer Schichten. Z. wiss. Photogr. **28**, 356—366 (1931).

AXELROD, D. J., HAMILTON, J. G.: Radio-autographic studies of the distribution of lewisite and mustard gas in skin and eye tissues. Amer. J. Path. **23**, 389—398 (1947).

BASERGA, R.: Autoradiographic methods. In H. BUSCH (ed.), Methods in cancer research, vol. I, 45—116. New York, London: Academic Press 1967.

BASERGA, R., MALAMUD, D.: Autoradiography. Techniques and application. New York, Evanston, and London: Harper & Row Publishers 1969.

BECKER, K.: Filmdosimetrie. Berlin, Göttingen, Heidelberg: Springer 1962.

BERLIN, M., ULLBERG, S.: Accumulation and retention of mercury in the mouse. I. An autoradiographic study after a single intravenous injection of mercuric chloride. Arch. environm. Hlth. **6**, 589—601 (1963).

Blackett, N. M., Kember, N. F., Lamerton, L. F.: The measurement of radiation dosage distribution by autoradiographic means with reference to the effect of bone-seeking isotopes. Lab. Invest. **8**, 171—178 (1959).

Bleecken, S.: Untersuchung des autoradiographischen Auflösungsvermögens mit Strahlungsquellen verschiedener Betaenergien. Atompraxis **9**, 321—324 (1961).

Boström, H., Odeblad, E., Friberg, U.: A quantitative autoradiographic study of the incorporation of S^{35} in tracheal cartilage. Arch. Biochem. **38**, 283—286 (1952).

Bothe, W.: Über photographische β-Strahlenmessung. Z. Phys. **8**, 243—250 (1922).

Branson, H., Hansborough, L. A.: The quantitative theory of autoradiography illustrated through experiments with P^{32} in chick embryo. Science **108**, 327—328 (1948).

Cobb, J., Solomon, A. K.: The detection of beta-radiation by photographic film. Rev. Sci. Instr. **19**, 441—447 (1948).

Dendy, P. P.: A method for the automatic estimation of grain densities in microautoradiography. Phys. in Med. y Biol. **5**, 131—137 (1960).

Dörmer, P.: Erfahrung mit der photometrischen Silberkornzählung in der Autoradiographie. Leitz Mitt. Wiss. Techn. **4**, 74—78 (1967a).

Dörmer, P.: Auflichtphotometrische Untersuchungen zur Größe der Koinzidenz in der Autoradiographie mit Tritium. Histochemie **8**, 1—8 (1967b).

Dörmer, P., Brinkmann, W.: Silberkornzählung mit dem Auflicht-Mikroskopphotometer. Ein Beitrag zur quantitativen Autoradiographie. Acta histochem., Suppl. VIII, 163—169 (1968).

Dörmer, P., Brinkmann, W.: Auflichtphotometrie von Mikroautoradiogrammen für quantitative Einbaustudien an Einzelzellen. Z. Anal. Chem. **252**, 84—89 (1970).

Dörmer, P., Brinkmann, W., Stieber, A., Stich, W.: Automatische Silberkornzählung in der Einzelzell-Autoradiographie. Eine neue photometrische Methode für die quantitative Autoradiographie. Klin. Wschr. **44**, 477—482 (1966).

Dörmer, P., Möller, E. D.: Autoradiography of the non-uniformity of cell kinetics as revealed in the forestomach of the mouse. Exp. Cell Res. **49**, 495—503 (1968).

Dörmer, P., Reichart, B., Huhn, D.: Kerntrockengewicht und Beteiligung von ^{3}H-Thymidin an der DNS-Synthese in Einzelzellen der regenerierenden Rattenleber. Z. Zellforsch. **86**, 559—570 (1968).

Domingues, F. J., Sarko, A., Baldwin, R. R.: A simplified method for quantitation of autoradiography. Int. J. appl. Radiat. **1**, 94—101 (1956).

Donaldson, P. E. K.: Electronic apparatus for biological research. London: Butterworths Scientific Publ. 1958.

Doniach, I., Pelc, S. R.: Autoradiograph technique. Brit. J. Radiol. **23**, 184—192 (1950).

Dudley, R. A.: Photographic detection and dosimetry of beta rays. Nucleonics **12**, 24—31 (1954).

Dudley, R. A., Dobyns, B. M.: The use of autoradiographs in the quantitative determination of radiation dosages from Ca^{45} in bone. Science **109**, 327—342 (1949).

Dudley, R. A., Pelc, S. R.: Automatic grain counter for assessing quantitatively high-resolution autoradiographs. Nature **172**, 992—993 (1953).

Eggert, J., Kuester, A.: Grain size determination and other applications of the callier effect. J. Soc. Mot. Pict. Eng. **30**, 181—190 (1938).

Engels, W.: Der zeitliche Ablauf von Protein- und Kohlenhydratsynthesen in der Oogenese bei Apis mellificia L. Verh. Dtsch. Zool. Ges. 243—251 (1965).

Evans, T. C.: Selection of radioautographic technique for problems in biology. Nucleonics **3**, 52—59 (1948).

Exner, G.: Über die Abhängigkeit der gemessenen Substanzmengen von der Fokussierung. Acta histochem. **9**, 166—168 (1960).

FICQ, A., ERRERA, M.: Analyse autoradiographique de la phénylalanine-2-^{14}C dans les noyaux isolés. Exp. Cell Res. **14**, 182—192 (1958).

FISCHER, H. A., WERNER, G.: Eine Methode zur quantitativen Auswertung von Autoradiogrammen auf der Grundlage der Durchlicht-Mikro-Photometrie. Histochemie **15**, 84—92 (1968).

FISCHER, H. A., WERNER, G.: Autoradiographie. Berlin: Walter de Gruyter & Co. 1971.

FRIESER, H., HEIMANN, G., RANZ, E.: Einwirkung radioaktiver Nuklide auf photographische Schichten. Phot. Korresp. **98**, 131—140 (1962).

GLOCKER, R.: Das photographische Schwärzungsgesetz für Elektronenstrahler verschiedener Energie. Z. Physik **160**, 568—572 (1960).

GROSS, J., BOGOROCH, R., NADLER, N. J., LEBLOND, C. P.: The theory and methods of the radioautographic localization of radioelements in tissues. Amer. J. Roentgenol. **65**, 420—458 (1951).

GULLBERG, J. E.: A new change-over optical system and a direct recording microscope for quantitative autoradiography. Exp. Cell Res., Suppl. **4**, 222—230 (1957).

HERRMANN, R.: Empfänger (SEV) und Anzeigegeräte: Formulierung der Anforderungen. Acta histochem. Suppl. VI, 189—201 (1965).

HERZ, R. H.: The photographic action of ionizing radiations. New York, London, Sydney, Toronto: John Wiley & Sons Inc. 1969.

JAMES, T. H., FORTMILLER, L. J.: Dependence of covering power and spectral absorption of developed silver on temperature and composition of the developer. Phot. Sci. Eng. **5**, 297—304 (1961).

JAMES, T. H., HIGGINS, G. C.: Fundamentals of photographic theory. New York: John Wiley & Sons, Inc. 1948.

JOWSEY, J.: Densitometry of photographic images. J. appl. Physiol. **21**, 309—312 (1966).

JOWSEY, J., LAFFERTY, W., RABINOWITZ, J.: Analysis of distribution of Ca^{45} in dog bone by a quantitative autoradiographic method. J. Bone Jt. Surg. **47**-A, 359—370 (1965).

KLEIN, E.: Die Beziehung zwischen der Schwärzung und der Größe der entwickelten Silberaggregate. Z. Elektrochem. **62**, 993—999 (1958).

KUTZIM, H.: Die quantitative Bestimmung der Verteilung von S^{35}-Sulfat bei der Maus mittels Autoradiographie. Nucl. Med. **3**, 39—50 (1963).

LÄUCHLI, A., LÜTTGE, U.: Untersuchung der Kinetik der Ionenaufnahme in das Zytoplasma von Mnium-Blattzellen mit Hilfe der Mikroautoradiographie und der Röntgen-Mikrosonde. Planta **83**, 80—98 (1968).

LAMERTON, L. F., HARRISS, E. B.: Resolution and sensitivity considerations in autoradiography. J. Phot. Sci. **2**, 135—144 (1954).

MAMUL, YA V.: Quantitative autoradiography using a radioactive wedge. Int. J. Appl. Radiat. **1**, 178—183 (1956).

MARINELLI, L. D., HILL, R. F.: Radioautography. Some physical and radiobiological aspects of the technique as applied to thin specimens. Amer. J. Roentgenol. **59**, 396—403 (1948).

MARSHALL, J. H., ROWLAND, R. E., JOWSEY, J.: Microscopic metabolism of calcium in bone. II. Quantitative autoradiography. Radiat. Res. **10**, 213—233 (1959).

MAURER, W., PRIMBSCH, E.: Größe der β-Selbstabsorption bei der ^{3}H-Autoradiographie. Exp. Cell Res. **33**, 8—18 (1964).

MAZIA, D., PLAUT, W. S., ELLIS, G. W.: A method for the quantitative assessment of autoradiographs. Exp. Cell Res. **9**, 305—312 (1955).

MERTZ, M.: Bestimmung der Silberkorngröße in Autoradiogrammen bei Auflicht und Durchlicht. Histochemie **17**, 128—137 (1969).

NADLER, N. J.: Some theoretical aspects of radioautography. Canad. J. Med. Sci. **29**, 182—194 (1951).

Nadler, N. J.: The quantitative estimation of radioactive isotopes by radioautography. Amer. J. Roentgenol. **70**, 814—823 (1953).

Nutting, P. G.: On the absorption of light in heterogeneous media. Phil. Mag. **26**, 423—426 (1913).

Odeblad, E.: Contributions to the theory and technique of quantitative autoradiography with P^{32} with special reference to the granulosa tissue of the Graafian follicles in the rabbit. Acta Radiol. Suppl. **93**, 1—123 (1952).

Odeblad, E.: Matrix theory for quantitative evaluation of autoradiographs. Lab. Invest. **8**, 113—126 (1959).

Pelc, S. R.: Quantitative aspects of autoradiography. Exp. Cell Res., Suppl. **4**, 231—237 (1957).

Pelc, S. R.: On the question of automatic or visual grain counting. Lab. Invest. **8**, 127—130 (1959).

Perry, R. P.: Quantitative autoradiography. In: D. M. Prescott (ed.), Methods in cell physiology, vol. I, p. 305—326. New York: Academic Press 1964.

Poddar, R. K.: On the quantitative relation between isotopic beta radiation and its photographic response. Indian J. Physiol. **29**, 189—198 (1955).

Preisser, G.: Stabilisierungsprobleme. Acta histochem. Suppl. VI, 165—169 (1965).

Ritzen, M.: Mast cells and 5-HT. Uptake of labelled 5-hydroxytryptamine (5-HT) and 5-hydroxytryptophan in relation to storage of 5-HT in individual rat mast cells. Acta physiol. scand. **69**, 1—12 (1967a).

Ritzen, M.: A method for the autoradiographic determination of absolute specific radioactivity in cells. Exp. Cell Res. **45**, 250—252 (1967b).

Rogers, A. W.: A simple photometric device for the quantitation of silver grains in autoradiographs of tissue sections. Exp. Cell Res. **24**, 228—239 (1961).

Rogers, A. W.: Die Mikroskopie von Autoradiographien. Leitz Mitt. Wiss. Techn. **3**, 43—47 (1964).

Rogers, A. W.: Techniques of autoradiography. Amsterdam, London, New York: Elsevier Publishing Company 1967.

Sandritter, W.: Über die Bedeutung des Fokussierungsfehlers in der Mikrophotometrie. Acta histochem. Suppl. VI, 203—205 (1965).

Schreiber, G., Lesch, R., Weinssen, U., Zähringer, J.: The distribution of albumin synthesis throughout the liver lobule. J. Cell Biol. **47**, 285—289 (1970).

Svedberg, T., Andersson, H.: On the relation between sensitiveness and size of grain in photographic emulsions. Phot. J. **61**, 325—332 (1921).

Taylor, J. H.: Autoradiography at the cellular level. In G. Oster and A. W. Pollister (eds.), Physical techniques in biological research, voll. III, 545—580. New York: Academic Press 1956.

Tolles, E.: Methods of automatic quantitation of microautoradiographs. Lab. Invest. **8**, 99—112 (1959).

Vaughan, J., Owen, M.: The use of autoradiography in the measurement of radiation doserate in rabbit bones following the administration of Sr^{90}. Lab. Invest. **8**, 181—191 (1959).

Waser, P. G., Lüthi, U.: Über die Fixierung von ^{14}C-Curarin in der Endplatte. Helv. Physiol. Acta **20**, 237—251 (1962).

Weber, K.: Leitz-Mikroskop-Photometer MPV mit variabler Meßblende. Leitz Mitt. Wiss. Techn. **3**, 103—107 (1965).

Westphal, A.: Einführung in die Reflexmikroskopie, 1—112. Stuttgart: Georg Thieme Verlag 1963.

Chapter 2

Quantitative Autoradiography in the Presence of Crossfire

J. B. Passioura
CSIRO, Division of Land Research
Canberra, Australia

Contents

2.1. *Introduction*

To make autoradiography quantitative is difficult. Apart from the many technical problems involved in preparing an acceptable autoradiograph, there are the analytical problems of:

(1) determining the distribution of density in the autoradiograph,
(2) relating the density to the exposure (that is, the radiation received),
(3) correcting the exposure for crossfire and self-absorption so as finally to determine the distribution of radioisotope in the section.

Much work has been done on problem (1); techniques of grain-counting and densitometry are fairly well developed (ROGERS, 1967). Problem (2) can be solved by calibrating the autoradiograph with another similar autoradiograph prepared from a series of standard sources. Problem (3) is, however, difficult. It has received little attention apart from the work of ODEBLAD (1956, 1957, 1959) and of BLEECKEN (1968a, b, c), whose approaches are rather difficult to apply in practice. The following discussion is concerned primarily with the problem of crossfire, that is, with the blurring which occurs because a point in the emulsion receives radiation from all points in the section. The problem of self-absorption is avoided by assuming that the distribution of isotope is uniform throughout the thickness of the section as is described in the theory below.

2.2. *Theory*

The geometric relation between the section and the emulsion is shown in Figure 2.1. It is assumed (i) that the section, the interspace, and the emulsion have uniform physical properties and, in particular, that they are

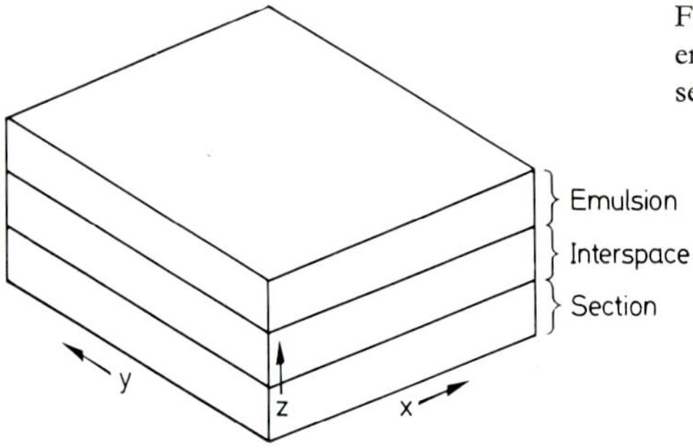

Fig. 2.1. Spatial relations between the emulsion, the interspace, and the section

uniformly thick, (ii) that the distribution of radioisotope in the section may be a function of x and y, but not of z. The first assumption should hold for well-prepared sections used with stripping film or other geometrically uniform emulsions; it does not hold if liquid emulsions are used. The second assumption should be adequately satisfied for thin sections; if it is not satisfied, the problem of correcting for crossfire becomes intractable using the present approach, although ODEBLAD's approach (1959) is, in principle, able to cope.

If the two assumptions hold, a line source of radioisotope at any position x, y in the section will produce an exposure of the emulsion which will depend only on the strength of the source and on a unique transfer function which is independent of x and y. The transfer function is defined as the exposure produced by a line source of unit strength. Thus, if we imagine the section to be made up of an infinite number of line sources of variable but known strength, we may, in principle, operate on these line sources with the transfer function, integrate the result, and so predict the exposure of the emulsion. Conversely, if we know the exposure of the emulsion and the transfer function, we can, in principle, calculate the distribution of radioisotope in the section. This argument is developed in detail below.

2.2.1. The One-Dimensional Autoradiograph

For a one-dimensional autoradiograph, that is, one in which the source strength varies with x but not with y, the relation between the exposure of the emulsion, $E(x)$, and the distribution of source strength in the section, $S(x)$, is given by the convolution integral

$$E(x) = \int_{-\infty}^{\infty} S(\eta)\, F(x-\eta)\, d\eta \qquad (2.1)$$

where $F(x-\eta)$ is the transfer function and η is a dummy variable which allows us to scan S after having fixed a value for x.

Equation (2.1) states that the exposure of the emulsion at any point x comprises the sum of the radiation received from all points in the section, and that the radiation received at x from any point η (distant $x-\eta$ from x) is given by the product of the source strength at η and the appropriate value of the transfer function. For convenience we assume that the system extends indefinitely in the x-direction and thus use infinite limits for the integral. E and S are not given units here, but E is defined as having a value of unity when produced by a section containing a uniformly distributed source of unit strength.

If we know $E(x)$ and $F(x-\eta)$ we can use equation (2.1) to calculate $S(x)$. $E(x)$ can be determined from the autoradiograph by densitometry or by counting grains. $F(x-\eta)$ can be found directly from an autoradiograph of a plane source at x (that is, a source distributed uniformly through a y, z plane of the section) but, in practice, it is more convenient to infer it from an extended source of known distribution. The simplest such distribution is the step function

$$\begin{aligned} S(x) &= S_0 \text{ (i.e. a constant) when } x \leqq 0 \\ S(x) &= 0 \quad \text{when} \quad x > 0\,. \end{aligned} \tag{2.2}$$

Combining (2.1) and (2.2), we have,

$$E(x) = S_0 \int_{-\infty}^{0} F(x-\eta)\, \mathrm{d}\eta \tag{2.3}$$

which, on putting $x-\eta=\xi$ and differentiating with respect to x, gives

$$-\frac{1}{S_0}\frac{\mathrm{d}E}{\mathrm{d}x} = F(\xi)\,. \tag{2.4}$$

Thus $F(\xi)$ is given by the negative slope of E/S_0, the normalized exposure.

An idealized example of equation (2.3) is given in Figure 2.2 where E/S_0 and S/S_0 are plotted as functions of x. The scale of x is chosen arbitrarily so that the main variation in E/S_0 and hence in F takes place over a distance of 5 units. Figure 2.3 gives F, the negative gradient of E/S_0.

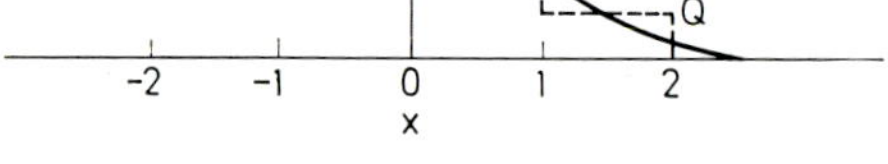

Fig. 2.2. E/S_0 and S/S_0 as functions of x when S is the step function described by Eq. (2)

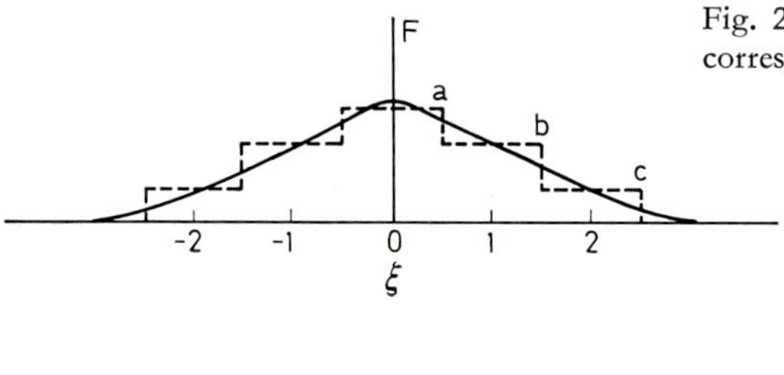

Fig. 2.3. The transfer function, F, corresponding to Fig. 2

It is usually more convenient to treat E/S_0 and F not as continuous functions but as series of steps (the dotted lines in Figures 2.2 and 2.3). When this is done, P and Q represent the average values of E/S_0 in the intervals $0 < x < 1$ and $1 < x < 2$ respectively, and a, b, and c represent the average values of F in the appropriate intervals of ξ. The exposure P arises because of crossfire from the region $x < 0$ of the section, and, in particular, comprise b units of radiation from the region $-1 < x < 0$ and c units from the region $-2 < x < -1$. Similarly, the exposure Q comprises c units of radiation from the region $-1 < x < 0$. Thus, having measured P and Q from the autoradiograph, and noting that the area under F is $1 = a + 2b + 2c$ (neglecting the small areas contained outside $\xi = \pm 2.5$), we have

$$a = 1 - 2P, \quad b = P - Q, \quad c = Q$$

which defines the transfer function.

This transfer function can now be used to calculate $S(x)$ using equation (2.1) in the form of a sum, namely,

$$E_{(i)} = a S_{(i)} + b (S_{(i+1)} + S_{(i-1)}) + c (S_{(i+2)} + S_{(i-2)}) \tag{2.5}$$

where i is the integral number of distance units from an arbitrary origin. It is possible to solve this equation using matrix algebra, but it is more convenient to solve it by successive approximations. A computer program for doing this (AUTRAD1) is given in the Appendix (2.4). The program is written in FORTRAN and consists essentially of the following steps:

1. Read in the distribution of exposure (E(I)) and the transfer function (A, B, C).
2. Make a first estimate of the distribution of source strength (S(I)) by equating it with E(I).
3. Operate on S(I) with A, B, and C to give G(I) the calculated exposure resulting from S(I).
4. Subtract G(I) from E(I) and add to S(I) to give an improved estimate of the source strength.
5. Repeat steps 3 and 4 until the difference between G(I) and E(I) is negligible. Twenty cycles should suffice. S(I) then gives the distribution of source strength.

2.2.2. The Two-Dimensional Autoradiograph

For a two-dimensional autoradiograph S varies with both x and y; F is a radially symmetrical function in two dimensions. Choosing the scale of x and y as before, so that the main variation in F takes place over five units, F as a step function may be depicted as in Figure 2.4. The values

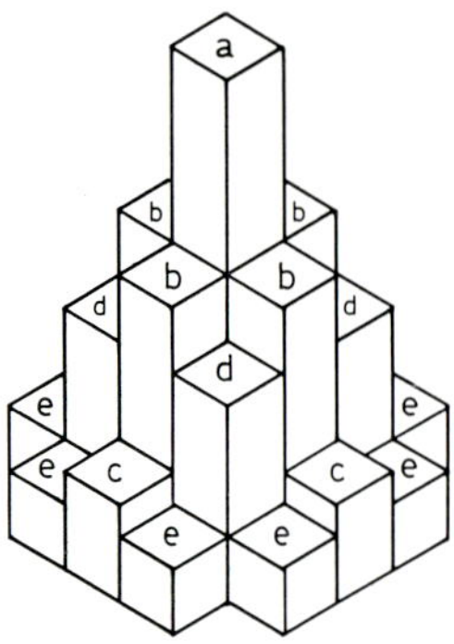

Fig. 2.4. The two-dimensional transfer function

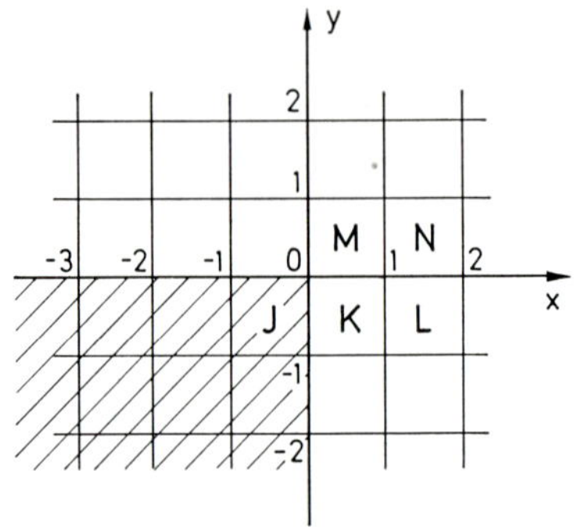

Fig. 2.5. Spatial relations between a source having the distribution described by Eq. (6) (the hatched region) and the exposures, J, K, L, M, and N, arising from the source. J, K, L, M, and N, represent average values of E/S_0 in the regions in which they appear

a, *b*, *c*, *d*, and *e*, may be defined directly by making an autoradiograph of a line source but, in practice, it is easier to infer them from an autoradiograph of a source having the distribution:

$$\begin{aligned} S(x,y) &= S_0, \quad \text{for } x \leqq 0 \text{ and } y \leqq 0 \\ S(x,y) &= 0, \quad \text{for } x > 0 \text{ or } y > 0. \end{aligned} \tag{2.6}$$

The exposure arising from such a source is illustrated in Figure 2.5. The normalised exposures K, L, M, and N arise from crossfire from the region $x \leqq 0$ and $y \leqq 0$. They are related to a, b, c, d, and e in the following way:

$$\begin{aligned} K &= b + c + d + 2e, \\ L &= \phantom{b +{}} c \phantom{{}+ d} + e, \\ M &= \phantom{b + c +{}} d + 2e, \\ N &= \phantom{b + c + d + 2{}} e, \end{aligned}$$

(that is: N receives e units of radiation from the region $-1 < x < 0$, $-1 < y < 0$; M receives d units from the region $-1 < x < 0$, $-1 < y < 0$ and e units from each of the regions $-2 < x < -1$, $-1 < y < 0$ and $-1 < x < 0$, $-2 < y < -1$; and so on). In addition, we have

$$1 = a + 4b + 4c + 4d + 8e.$$

Thus,

$$a = 1 - 4K,\quad b = K - L - M + N,\quad c = L - N,\quad d = M - 2N, e = N.$$

Alternatively, for a, we have $a = J - 2K + M$, which should be more accurate.

The transfer function is defined by a, b, c, d, and e, and can be used to calculate $S(x,y)$ from $E(x,y)$ by successive approximations as for the one-dimensional system. A computer program for doing this (AUTRAD2) is given in the Appendix (2.4). The program is essentially the same as AUTRAD1. The only major difference (apart from the dimensions) is that the second partial derivatives are subdued by averaging (statement 92 and the four following it) if they are numerically greater than the first partial derivatives. This has been found necessary because without it, unlike AUTRAD1, AUTRAD2 does not converge reliably, and small errors in $E(x,y)$ can result in wild deviations in $S(x,y)$.

2.3. *Practice*

There are several practical implications and difficulties arising from the above theory.

2.3.1. The Transfer Function

The transfer function describes empirically the several aspects of autoradiography which ODEBLAD (1959) and BLEECKEN (1968a, b, c) consider in detail: all of the effects of geometry, of absorption, and of scatter, are lumped together into one statement. It is therefore important, when making an autoradiograph to determine a transfer function, that the setup be a faithful copy of that used when autoradiographs are made of experimental sections. Thus the emulsion, the conditions of development, the interspace, and the thickness and specific density of the section, must be identical in the two cases.

As described previously, a transfer function can be determined from an autoradiograph of a section containing an extended source of known distribution. To make such a section may be difficult, but this problem must be solved by the user working with his own materials. The author's autoradiographic work has been concerned with determining the distribution of isotope as a function of radial distance (up to 2 mm) from a plant root growing in synthetic, easily-sectioned soil. With such a system it is easy to make a suitable section to determine the transfer function. A section of uniformly labeled soil can be cut to the desired shape and matched with a complementary section of unlabeled soil. When working with cellular material at a microscopic level, however, the production of uniformaly labeled material and the cutting of sections to the desired shape may be difficult.

Both AUTRAD 1 and AUTRAD 2 assume that the transfer function has a diameter of five units. Because of the many sources of error involved in quantitative autoradiography, this level of detail would seem to be the best that can be hoped for. However, for extremely reliable data it may be worthwhile dividing the transfer function into seven units; this would require small modifications of the programs.

2.3.2. One-Dimensional or Two-Dimensional?

AUTRAD 1 is much easier to use than AUTRAD 2. It requires many fewer measurements of exposure, it converges more reliably, and the transfer function is easier to determine. But to use it, one must be sure that the part of the autoradiograph being measured is effectively one-dimensional. In practice this means that the radius of curvature of iso-exposure lines must be greater than the effective diameter of the transfer function (five units in our arbitrary scale). For then, if the exposure for AUTRAD 1 is measured along a transect normal to the iso-exposure lines, AUTRAD 1 and AUTRAD 2 will give the same result.

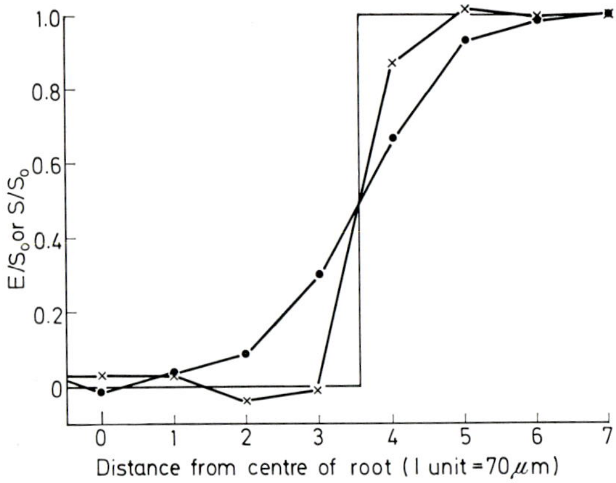

Fig. 2.6. Measured E/S_0 (●–●), estimated S/S_0(x—x), and true S/S_0 (solid line), as functions of distance, for an autoradiograph of a section of soil through which a root had been growing

2.3.3. An Example

Figure 2.6 shows the transformation which AUTORAD2 may make of experimental data. The measured exposures were of a radially symmetrical autoradiograph of a 50 μ section of soil through which a root of 500 μm diameter had been growing. The root had been removed before the section was made. The exposures shown were the means of several transects through the centre of the hole left by the root. The soil solution contained only $^{45}CaSO_4$ and conditions were such that the root had no influence on the distribution of Ca. This distribution was therefore known and is depicted as the thick line in Figure 2.4. The transfer function was: $a = 0.145$, $b = 0.080$, $c = 0.030$, $d = 0.055$, $e = 0.024$. The sharpening of the data by the program is considerable.

References

Bleecken, S.: Die Abbildungseigenschaften autoradiographischer Systeme. I. Grundlagen zur Berechnung der autoradiographischen Übertragung. Z. Naturforsch. **23**b, 1339—1349 (1968a).

Bleecken, S.: Die Abbildungseigenschaften autoradiographischer Systeme. II. Lineare und nichtlineare autoradiographische Übertragung. Z. Naturforsch. **23**b, 1350—1359 (1968b).

Bleecken, S.: Die Abbildungseigenschaften autoradiographischer Systeme. III. Autoradiographisches Auflösungsvermögen. Z. Naturforsch. **23**b, 1478—1495 (1968c).

Odeblad, E.: Matrix theory of quantitative apposition autoradiography. Acta Radiol. **45**, 323—339 (1956).

Odeblad, E.: Further approximate studies on beta ray absorption and transmission. Acta Radiol. **48**, 289—306 (1957).

Odeblad, E.: Matrix theory for quantitative evaluation of autoradiographs. Lab. Invest. **8**, 113—126 (1959).

Rogers, A. W.: Techniques of autoradiography. Amsterdam-London-New York: Elsevier 1967.

2.5. *Appendix: The Computer Programs*

Note 1. The programs are written in a fairly advanced form of FORTRAN in order to save space. Consequently, they may need minor modifications to run on some computers.

Note 2. AUTORAD1 operates on E(I), S(I) and G(I) only from I = 3 to I = N — 2. Consequently, E(1) and E(2) should be set equal to E(3), and E(N) and E(N — 1) should be set equal to E(N — 2). Similar remarks apply to AUTORAD2 for both I and J.

```
      PROGRAM AUTRAD1
C     THIS PROGRAM CORRECTS THE MEASURED EXPOSURE OF AN AUTORADIOGRAPH
C     SO AS TO DECREASE ERRORS ARISING FROM CROSSFIRE
C     THE AUTORADIOGRAPH IS ONE-DIMENSIONAL
C
C     E(I) IS THE EXPOSURE
C     S(I) IS THE ESTIMATED SOURCE STRENGTH
C     G(I) IS THE EXPOSURE CALCULATED FROM S(I)
C     M    IS THE NUMBER OF VALUES OF E(I)
C     A,B,C ARE THE TRANSFER FUNCTION
C     NCYCLES IS THE NUMBER OF CYCLES
C
      DIMENSION E(100),S(100),G(100)
    1 READ(60,2) M,NCYCLES
    2 FORMAT(2I10)
      IF(EOF,60)100,3
    3 READ(60,4) A,B,C
    4 FORMAT(3F10.3)
      WRITE(61,104) A,B,C
  104 FORMAT(///10X,8HTRANSFER,8HFUNCTION,/7X,3(2X,F10.3))
      READ(60,40)(E(I),I = 1,M)
   40 FORMAT(16F5.2)
      WRITE(61,105)(E(I),I = 1,M)
  105 FORMAT(/10X,8HEXPOSURE,/(7X,10(2X,F10.3)))
      M2 = M-2
      K = 0
      DO 5 I = 1,M
    5 S(I)=E(I)
   12 DO 6 I = 3,M2
    6 G(I)=C*(S(I-2)+S(I+2))+B*(S(I-1)+S(I+1))+A*S(I)
      K = K+1
      DO 11 I = 3,M2
   11 S(I)=S(I)-(G(I)-E(I))
      IF(K.GT.NCYCLES)9,12
    9 WRITE(61,13)(S(I),I = 1,M)
   13 FORMAT(/10X,8HSOURCE S,7HTRENGTH,/(7X,10(2X,F10,3)))
      WRITE (61,15) K
   15 FORMAT(/10X,8HCYCLES = ,I4)
      GO TO 1
  100 END
```

```
      PROGRAM AUTRAD2
C     THIS PROGRAM CORRECTS THE MEASURED EXPOSURE OF AN AUTORADIOGRAPH
C     SO AS TO DECREASE ERRORS ARISING FROM CROSSFIRE
C     THE AUTORADIOGRAPH IS TWO-DIMENSIONAL
C      E(I,J) IS THE EXPOSURE
C      S(I,J) IS THE ESTIMATED SOURCE STRENGTH
C      SNEW(I,J) IS THE ESTIMATED SOURCE STRENGTH AFTER THE
C                SECOND DERIVATIVES HAVE BEEN SUBDUED
C      G(I,J) IS THE EXPOSURE CALCULATED FROM S(I,J)
C      M,N ARE THE NUMBER OF ROWS AND COLUMNS RESPECTIVELY OF E(I,J)
C      A,B,C,D,EE ARE THE TRANSFER FUNCTION
C      NCYCLES IS THE NUMBER OF CYCLES
      DIMENSION E(50,50),S(50,50),G(50,50),SNEW(50,50)
    1 READ(60,100) M,N,NCYCLES
  100 FORMAT(3I10)
      IF(EOF,60)10,2
    2 READ(60,101) A,B,C,D,EE
  101 FORMAT(16F5.0)
      WRITE(61,102) A,B,C,D,EE
  102 FORMAT(///5X,8HTRANSFER,8HFUNCTION,5F10.3//5X,8HEXPOSURE)
      DO 13 I = 1,M
   13 READ(60,101) (E(I,J),J = 1,N)
      DO 11 I = 1,M
   11 WRITE(61,103) (E(I,J),J = 1,N)
  103 FORMAT(16(1X,F7.3))
      M2 = M-2 $ N2 = N-2 $ K = 0
      DO 3 I = 1,M $ DO 3 J = 1,N
    3 S(I,J) = E(I,J)
    4 DO 5 I = 3,M2 $ DO 5 J = 3,N2
    5 G(I,J) = C*(S(I-2,J)+S(I+2,J)+S(I,J-2)+S(I,J+2)) + B*(S(I-1,J)+
     C     S(I+1,J)+S(I,J-1)+S(I,J+1))+A*S(I,J)+EE*(S(I-2,J-1)+S(I-2,J+1)
     C   +S(I-1,J+2)+S(I-1,J-2)+S(I+1,J-2)+S(I+1,J+2)+S(I+2,J-1) +
     C     S(I+2,J+1))+D*(S(I-1,J-1)+S(I+1,J-1)+S(I-1,J+1)+S(I+1,J+1))
      K = K+1
      DO 8 I = 3,M2 $ DO 8 J = 3,N2
    8 S(I,J) = S(I,J)-G(I,J)/2. +E(I,J)/2.
      DO 90 I = 1,M  $ DO 90 J = 1,N
      SNEW(I,J) = S(I,J)
      IF (I.LT.3.OR.I.GT.M2.OR.J.LT.3.OR.J.GT.M2) GO TO 93
      IF(ABS(S(I,J-1)-2.*S(I,J)+S(I,J+1)).LT.ABS(S(I,J-1) -S(I,J+1)))
     CGO TO 93
   91 IF(ABS(S(I-1,J) -2.*S(I,J)+S(I+1,J)).LT.ABS(S(I-1,J) -S(I+1,J)))
     CGO TO 93
   92 SN = S(I-1,J)
      IF(ABS(S(I-1,J) -S(I,J)).LT.ABS(S(I+1,J) -S(I,J))) SN = S(1+1,J)
      IF(ABS(S(I+1,J) -S(I,J)).LT.ABS(S(I,J-1) -S(I,J))) SN = S(I,J-1)
      IF(ABS(S(I,J-1) -S(I,J)).LT.ABS(S(I,J+1) -S(I,J))) SN = S(I,J+1)
      SNEW(I,J) = (S(I,J)+S(I-1,J)+S(I+1,J)+S(I,J-1)+S(I,J+1) -SN)/4.
   93 CONTINUE
   90 CONTINUE
      DO 94 I = 1,M $ DO 94 J = 1,N
   94 S(I,J) = SNEW(I,J)
      IF(K.LE.NCYCLES) GO TO 4
    9 WRITE(61,104)
  104 FORMAT(/5X,8HSOURCE S,7HTRENGTH)
      DO 12 I = 1,M
   12 WRITE(61,105) (S(I,J), J = 1,N)
  105 FORMAT(16(1X,F7.3))
      WRITE(61,106) K   $ GO TO 1
  106 FORMAT(10X,8HCYCLES = ,I4)
   10 END
```

Chapter 3

Microautoradiography of Water-Soluble Inorganic Ions

ULRICH LÜTTGE
Fachbereich Biologie - Botanik - der Technischen Hochschule Darmstadt, W.-Germany

Contents

3.1. *Introduction*

The major field of application of microautoradiography of inorganic compounds in the plant sciences has been ion transport-physiology. The first successful demonstration of an *in situ* distribution of water-soluble inorganic ions in a plant tissue known to the author was achieved during studies of ion translocation in the phloem (BIDDULPH, 1956).

This chapter describes techniques developed in the connection with ion transport studies, i.e. long distance transport of ions in the phloem and in the xylem (e.g. S. F. BIDDULPH, 1956; RINGOET et al., 1968; STÖSSER, 1970), ion transport in roots (e.g. WEIGL und LÜTTGE, 1962, 1965; LÜTTGE and WEIGL, 1962, 1964; BRANTON and JACOBSON, 1962; ZIEGLER et al., 1963; LÜTTGE, 1964b; GAHAN and RAJAN, 1965; GIELINK et al., 1966; S. F. BIDDULPH, 1967; KRICHBAUM et al., 1967; BARBER et al., 1968; CLARKSON et al., 1968; CLARKSON and SANDERSON, 1969), ion transport in leaves (e.g. BRANTON and JACOBSON, 1962; RINGOET et al., 1971a; LÄUCHLI und LÜTTGE, 1968), ion excretion and resorption by glands (e.g. LÜTTGE, 1964a, 1965, 1966; ZIEGLER and LÜTTGE, 1967; OSMOND et al., 1969; LÜTTGE et al., 1971), and ion distribution within individual plant cells (e.g. ZIEGLER und LÜTTGE, 1967; RECHENMANN et al., 1969; RINGOET et al., 1971b).

Hence the particular experimental requirements of translocation studies are reflected in the individual techniques described, although most of the procedures will be more generally applicable.

3.2. *The Choice of the Radioisotopes*

An evaluation of the basic principles of resolution in microautoradiography clearly shows that resolution decreases considerably with increasing energy, i.e. with increasing range of penetration of the radiation emitted by the radioisotope (BOYD, 1955; ROGERS, 1967). Therefore, the use of isotopes with a low energy β-radiation has been most successful in microautoradiography of biological specimens.

A selection of radioisotopes is listed in Table 3.1. The principal elements of organic compounds, i.e. H and C, have radioisotopes (i.e. ^{3}H and ^{14}C) emitting a radiation very suitable for microautoradiography; and ^{35}S and ^{33}P are yet two other appropriate radioisotopes of elements which are important in many organic substances. This fortunate coincidence has greatly facilitated the use of microautoradiography in the investigation of the distribution of organic compounds in biological specimens (see Chapters 4, 5, 6).

Table 3.1. Nuclides used in micro-autoradiography of water soluble inorganic ions.

Suitability for micro-autoradio-graphy	Radio-iso-tope	Radiation[a] β[b] %[e]	Energy[f] MeV	γ %[e]	Energy[f] MeV	X-rays %[e]	$t_{1/2}$[c]	Specific activity available[d]
autoradiography of organic compounds	^{3}H	100	0.018	—			12.3 yr	
	^{14}C	100	0.156	—			5730 yr	
autoradiography of inorganic anions	^{35}S	100	0.167	—			87 d	$^{35}SO_4^{--}$: carrier free
	^{33}P	100	0.248	—			25.2 d	$^{33}PO_4$: carrier free
	^{32}P	100	1.71	—			14.3 d	$^{32}PO_4$: 50—140 Ci/mg P
	^{36}Cl	98.3	0.71	—		1.7	3×10^5 yr	$^{36}Cl^-$ > 3 mCi/g Cl
autoradiography of inorganic cations	^{45}Ca	100	0.256	—			165 d	$^{45}Ca^{++}$: 10—25 mCi/mg Ca
	^{59}Fe	46	0.27	56	1.10		45 d	$^{59}Fe^{+++}$: 3—30 mCi/mg Fe
		53	0.46	44	1.29			
		plus 2 others		plus 3 others				
	^{86}Rb	91	1.77	9	1.08		18.7 d	$^{86}Rb^+$: 2—10 mCi/mg Rb
		9	0.69					
suitability for micro-autoradio-graphy limited	^{22}Na	89.7	0.54	99.94	1.28	10.2	2.62 yr	$^{22}Na^+$: 100 mCi/mg Na
		0.06	1.83					
	^{42}K	82	3.54		largely		12.4 hr	$^{42}K^+$: 250 mCi/g K
		18	1.98	18	1.52			

[a] Extracted from SEELMANN-EGGEBERT and PFENNIG (1964).

[b] β^--particles except for ^{22}Na where β^+-particles are emitted.

[c] $t_{1/2}$ = half life; hr = hour; d = day; y = year.

[d] Data were taken from suppliers catalogues.

[e] Percentages are based on the number of disintegrations in the parent radio-nuclide = 100%. Different γ-rays may be emitted in coincidence so that percentages > 100 are possible.

[f] MeV = million electron volts = 10^6 eV.

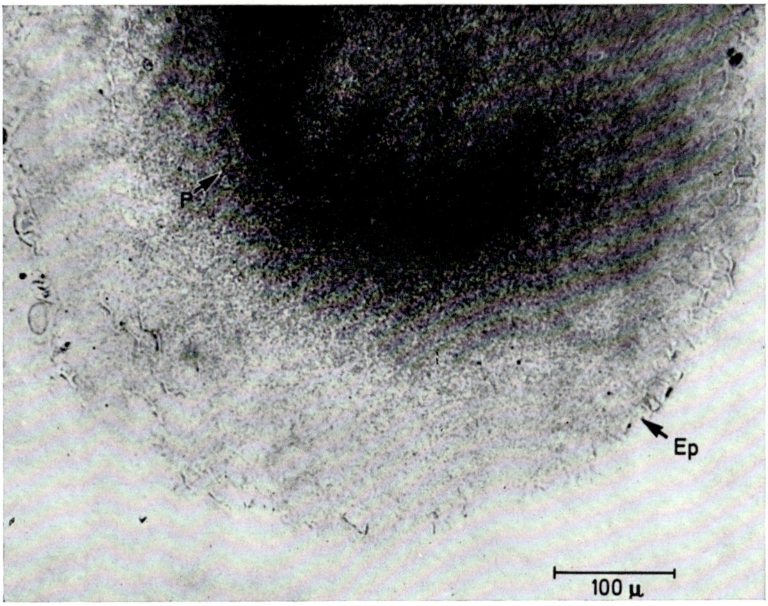

Fig. 3.1. ^{22}Na localization in a cross section of a maize root 5 cm from the root apex. Cryostat sectioning technique. Section thickness 12 μm; exposure 26 days. Ep = epidermis, P = pericycle. (With kind permission from SHONE et al., 1969)

3.2.1. Inorganic Cations

By contrast, among the most important ions regulating ionic relations of plant cells, i.e. Na^+, K^+ and Cl^-, only the latter has a radioisotope emitting β-particles of an energy suitable for microautoradiography and having an appropriate half-life. The radioisotopes of sodium and potassium are difficult to use because of their short half lifes and/or high energy β- and γ-radiations. Microautoradiographic investigations using $^{22}Na^+$ and $^{42}K^+$ have been tried occasionally, but the results are not convincing. The high radiation energies cause low resolution, and the handling of the short half-life ^{42}K in microautoradiography requiring extended times for preparation is rather awkward. Figure 3.1 shows that under favourable conditions gross distribution of sodium in different tissues of a plant organ can be elucidated using ^{22}Na-autoradiography (SHONE et al., 1969; CLARKSON and SANDERSON, 1970). However microautoradiography for ^{22}Na-localization on the cellular level so far has not yielded convincing results.

In a large number of non-autoradiographic transport-studies in plants $^{86}Rb^+$ has been used to label K^+-solutions, because Rb^+ is to a great extent transported in the same way as the potassium ion (cf. JESCHKE, 1970). The

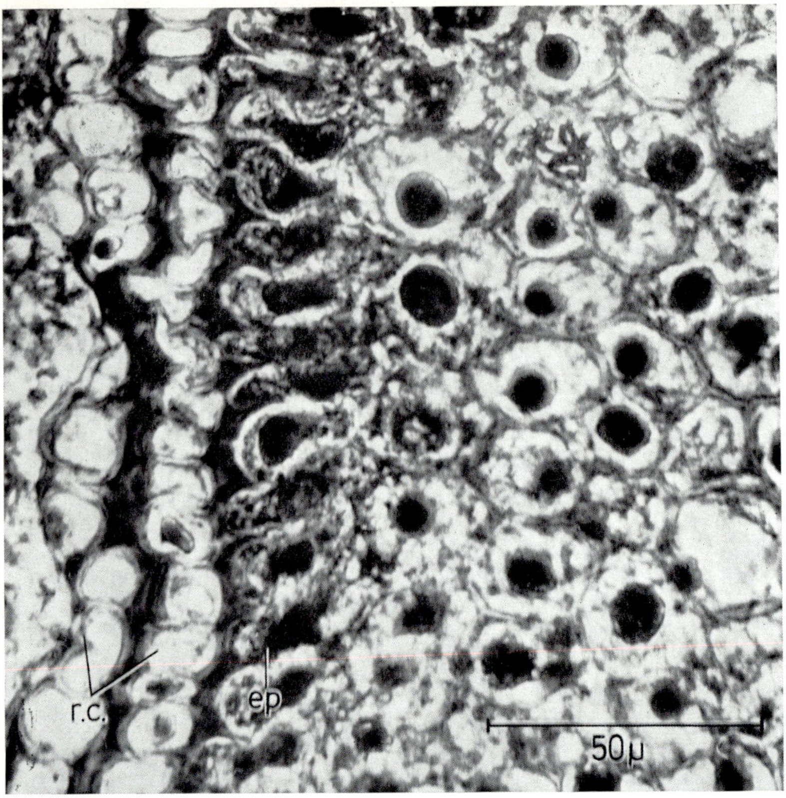

Fig. 3.2. The distribution of ^{46}Sc in a transverse section of an onion root 1.8 mm from the tip. Uptake period 30 min in 0.1 mM $^{46}ScCl_3$ (spec. activity 20 mCi/mM Sc^{3+}). (a) Stained cryostat section (Erlich's haematoxylin) showing cytological details: r. c. =

radiation emitted by ^{86}Rb is not ideal for autoradiography. However, in view of successful experiments with ^{32}P, which has a β-radiation similar in energy to the β-particles emitted by ^{86}Rb, there is at least a chance that ^{86}Rb could yield meaningful microautoradiographs. If experiments can be adjusted such, that the exposure time of autoradiographs is kept short, the disturbingly low resolution due to the γ-radiation would be minimized, because the amount of γ-radiation absorbed in a short period in the thin film will be much smaller than that of the β-particles. The successful use of $^{46}Sc^{+++}$ in microautoradiography of plant roots (CLARKSON and SANDERSON, 1969; Fig. 3.2; see below) emphasizes this point. Yet to the author's knowledge no extensive microautoradiographic work using ^{86}Rb has been performed with plant material.

In conclusion, microautoradiography of monovalent cations proves extremely difficult if not impossible because of the lack of suitable radio-

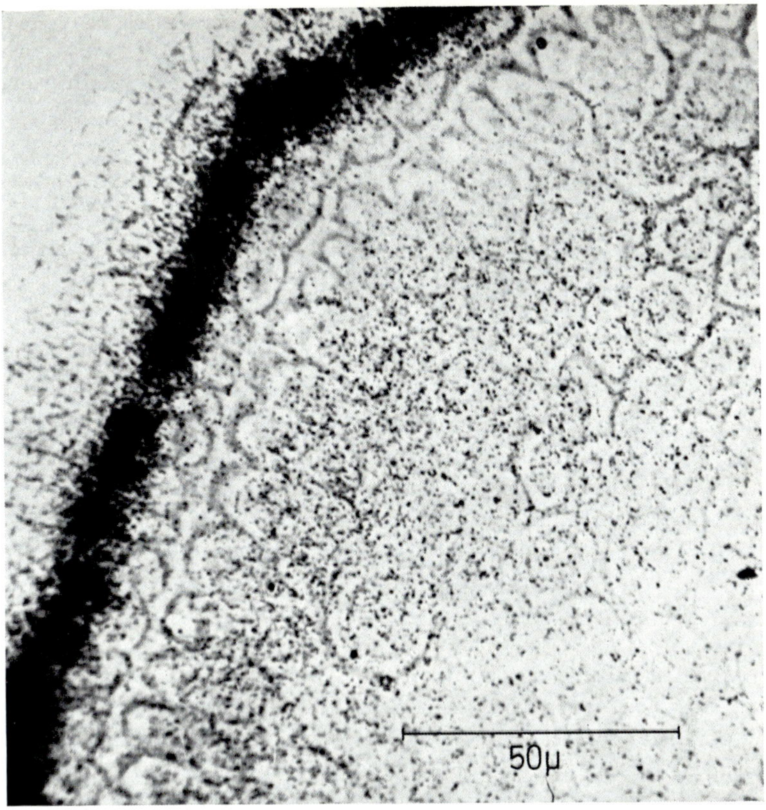

root cap, ep = epidermis. (b) Autoradiogram. Cryostat sectioning technique; Kodak AR. 10 stripping film; section thickness 20 µm; exposure 75 days at —25 °C. (With kind permission from CLARKSON and SANDERSON, 1969)

isotopes. Therefore, an important aspect of ion transport-physiology is not accessable by the microautoradiographic technique.

Among the polyvalent cations Ca^{++} has an isotope ideal for microautoradiography (^{45}Ca). Although the role of Ca^{++} in plant tissues is quite different from that of Na^{+} and K^{+}, many interesting findings of cation-transport have been made by ^{45}Ca-autoradiography (e.g. LÜTTGE und WEIGL, 1962; GIELINK et al., 1966; BIDDULPH, 1967; KRICHBAUM et al., 1967; RINGOET and DE ZEEUW, 1968; RINGOET et al., 1968; STÖSSER, 1970; RINGOET et al., 1971 a, b). ^{59}Fe is another radioisotope which has been successfully used in microautoradiographic studies (BRANTON and JACOBSON, 1962).

Of course, if one is interested in the study of transport and distribution of cations less common than those listed in Table 3.1, one will certainly find other useful radioisotopes in the chart of nuclides. Thus microautoradiography of Scandium-46 in plant roots has supplied quite reasonable autoradiographs of medium resolution (CLARKSON and SANDERSON, 1969; Fig. 3.2). The radioisotope data of ^{46}Sc are as follows

$$
\begin{array}{rrl}
\beta^-: & \sim 100\%, & 0.36\ \mathrm{MeV}, \\
 & 0.004\%, & 1.48\ \mathrm{MeV}, \\
\gamma: & 100\%, & 0.89\ \mathrm{MeV}, \\
 & 91\%, & 1.12\ \mathrm{MeV}, \\
t\,1/2: & 84\ \text{days}. &
\end{array}
$$

In spite of the γ-radiation present and the exposure times of up to 75 days resolution was very reasonable in these experiments. The precision of ^{46}Sc localization is probably facilitated by the fact that little of the $^{46}Sc^{+++}$ is actually in a diffusible condition in the cells, but is adsorbed on exchange-surfaces so that problems associated with autoradiography of water soluble substances are minimized (CLARKSON, personal communication).

3.2.2. Inorganic Anions

As already mentioned above, Cl^- is an important anion in plant cells and it has a radioisotope with a β-emission and a half-life very suitable for microautoradiography. A disadvantage of ^{36}Cl is that it is rather expensive. Furthermore, the fact that ^{36}Cl is only available at low specific activities will often make microautoradiographic investigations with this isotope difficult (see also Chapter 3.3.1.).

Hence $^{35}SO_4^{--}$ has often been preferred in microautoradiographic studies of anion transport. The physical properties of ^{35}S (radiation-energy and half-life) make it an ideal isotope for microautoradiography. Furthermore it is inexpensive, and it is available carrier free. The same is true for ^{32}P which, however, has a considerably higher β-radiation energy. Nevertheless, very good ^{32}P-microautoradiographs have been published. Of course, if ion transport *per se* is to be studied using $^{35}SO_4^{--}$ and $^{32}PO_4^{---}$ as labeled ions, one has to observe the appropriate controls taking into account metabolic incorporation of ^{35}S and ^{32}P into various organic compounds.

3.3. *The Design of the Experiments*

Basic experimental requirements are determined by the nature of the problem being investigated, and as with other techniques, much will depend on the investigator's intuition to decide whether autoradiography will be the appropriate tool to use. Furthermore the experimenter's experience and familiarity with the problem in question will allow him to select various modifications of experimental design and thus to approach a solution of his problem. In this sense, very little can be said in general on the design of microautoradiographic experiments. However, there are two particular questions on which a few general remarks should be made, i.e.:

(1) The dosage of radioactivity and exposure time required to obtain reasonable silver grain densities on the microautoradiographs, and
(2) the interpretation of static autoradiographic images in relation to dynamic events such as transport mechanisms.

3.3.1. Exposure Time and Specific Activities of External Solutions as Important Factors for Silver Grain Yield and their Relation to Methods Chosen for Data Documentation

The first question here is: what is a reasonable grain density? The answer will depend to some extent on the chosen method of documentation of the results (see also Chapter 1). If the results are quantitatively documented by grain counting, grain density in the developed autoradiographs should not be too high, so that grain numbers in the individual squares of the microscopic eye piece-grid used can be easily detected visually. If quantitative photometry of the autoradiographs is intended, the grain density must be within the limits of proportionality between radioactivity in the tissue and the grain numbers and optical signals obtained (see Chapter 1). For semiquantitative or non-quantitative documentation, e.g. by micro-photography very high grain densities may occasionally be an advantage (Fig. 3.3).

The grain density on the autoradiograph for a given radioisotope used is dependent on a variety of factors, e.g. on

- the specific activity of the labeled experimental solution;
- the rate of uptake, incorporation or accumulation of the labeled compound or ion used;
- the distribution of the labeled compound or ion within the tissue investigated (extremes being equal distribution in the whole tissue and concentration in particular cells or organelles respectively);
- the section thickness;
- the sensitivity of the photographic emulsion chosen;
- the exposure time;
- the time of photographic development.

Most of these factors are interdependent. For instance when the rate of uptake of a labeled ion from an external solution is small, a high specific activity in the external solution may be required to introduce a sufficient

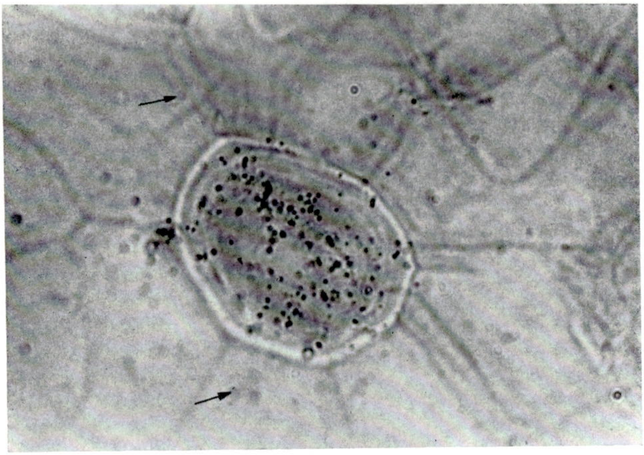

a

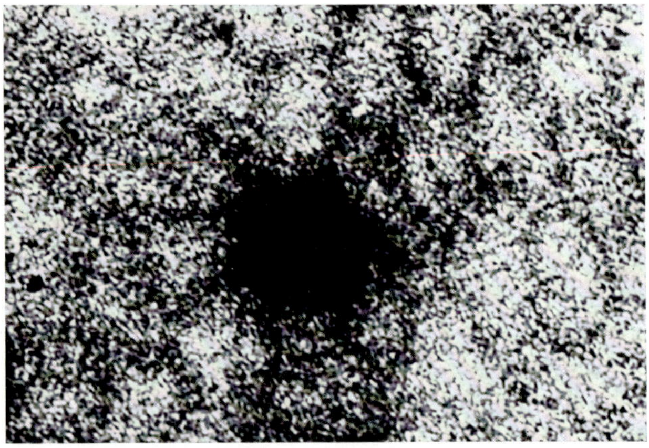

b

Fig. 3.3. Micro-autoradiographs of hydropote glands of *Nymphaea* leaves after $^{35}SO_4^{--}$-uptake. The hydropotes have a circular circumference in the face on views shown. (a) Grain yield adequate for grain counting. The autoradiogram is in contact with the tissue; grains over the epidermal cells (→) are not in focus. (b) Grain yield inadequate for grain counting and above the hydropotes also inadequate for microdensitometry. However, the distribution of silver grains in the epidermal cells is much clearer on this overexposed autoradiogram than in (a). The autoradiogram has been separated from the tissue, and all grains are in focus. Magnifications: (a) 900 ×, (b) 500 ×. [(a): from LÜTTGE 1964; (b): original]

amount of radioactivity into the tissue sampled for microautoradiography. Less radioactivity will be necessary in the uptake solution, if the labeled ions taken up are concentrated in a given compartment rather than distributed equally within the tissue. With increasing section thickness grain yield on the autoradiograph will increase too, but resolution will decrease. An extension of exposure time will also raise the amount of silver grains in the film as more radiation is absorbed. However, it is useless to extend exposure times for more than 3 half-lives at the most, because this will reduce the radioactivity present in the tissue to 12.5% of the amount initially taken up. Extended exposure times also lead to reduced resolution. Furthermore, since microautoradiographic emulsions do not generally have the high stability of the latent photographic image which is characteristic for most ordinary commercial films, very long exposure times will increase the probability of latent image fading. To extend time of the development may be useful as it increases the speed of the photographic emulsion. However, this will be accompanied by a loss of resolution as grain sizes are increased with increasing development times.

In view of the interactions among the many factors which may be commanded by the experimenter to determine the autoradiographic image, the best advice that can be given is to find the appropriate parameters by trial and error, varying some of these factors within reasonable limits. The study of SO_4^{--} uptake into the hydropote gland cells of water lily leaves (*Nymphaea*) may illustrate this view (Fig. 3.4, 3.5, Table 3.2, LÜTTGE, 1964; LÜTTGE et al., 1971).

In these experiments a number of the factors listed above were given as follows:

The radioisotope used was ^{35}S; the period of uptake was 3 hrs, since preliminary experiments had shown that within this period SO_4^{--} uptake into the tissue was linear with time (cf. Fig. 3.6); photographic emulsion was Kodak AR 10 stripping film; photographic development time was 5 min in Kodak D 19b developer at 18°C. The aim of the experiments was to determine the dependence of SO_4^{--}-uptake on SO_4^{--}-concentration of the external solution. Parameters varied were: specific activity of the external solutions, exposure time and the method of quantitising the results. For preparation of the autoradiographs a special procedure was used as described in detail in Chapter 3.4.2.6.1.

Quantitative measurements were made by microdensitometry as well as by grain counting. In microdensitometry (index "d" stands for "densitometry") the extinction ($\log_{10} B/A$) was calculated from transmission of the light by a blank part (B) of the film (i.e. where the film was not in contact with the tissue) and from the transmission of the autoradiographs (A). Extinction is proportional to the amount of label present in the section measured. Measurements of A and B were taken using oil immersion lenses with 75× and 100× magnification respectively; the measured areas were circles of a diameter of 6.8 µm and 5.1 µm respectively. The extinctions were divided by the specific activity of the external solution S [µCi/µmole] and thus relative units for SO_4^{--} uptake were obtained:

$$\text{Uptake} = \frac{1}{S} \log_{10} \frac{B}{A} \left[\frac{\text{extinction}}{\mu\text{Ci}/\mu\text{mole}} \equiv F_d \times \mu\text{mole} \right] \tag{3.1}$$

where F_d is a factor of proportionality which does not depend on the specific activity of the external solution but is determined by parameters such as exposure time, conditions of photographic development and others (see above) which affect grain yield and size on the autoradiographs. Similarly in grain counts (index "c" stands for "counting") the number of grains (z) per given area of the autoradiograph [e.g. in the experiments described here (5.1 μm^2) and (6.7 μm^2) at the magnifications 10×100 and 10×75 respectively], minus the number of background grains (z′), divided by the specific activity is proportional to μmoles accumulated in the counted area of the tissue section:

$$\text{Uptake} = \frac{(z - z')}{S} \times \left[\frac{\text{number of grains}}{\mu m^2 \cdot \mu Ci/\mu mole} \equiv F_c \times \mu mole\right] \qquad (3.2)$$

F_c, like F_d above, is a factor of proportionality dependent on various parameters but independent of specific activity. Attempts to eliminate the factors F_d in microdensitometry and F_c in grain-counting by taking measurements on autoradiographs of standards (see Chapter 1) have not been made. However, the factors were constant for any given experiment where all slides were treated simultaneously (see Chapter 1). With the various factors kept constant, as indicated above, the results obtained according to Eqs. (3.1) and (3.2) should only depend on exposure time, i.e. factors F_d and F_c respectively should be similar for different experiments where exposure time is the same.

The tabulation of Fig. 3.4 shows the experimental details of 3 experiments. Exposure time was varied in such a way, that at various intervals slides were developed. An exposure time of about 20 days turned out to be most appropriate. Results obtained for shorter exposure times are not shown. Thus exposure time was very similar in the 3 experiments depicted. The external ion concentration was varied between 0.01 and 0.5 mM K_2SO_4 and the specific activity was between 5900 and 5 mCi/mM SO_4^{--}. (Specific activities which did not yield reasonable autoradiographs are not listed.)

It can be seen that with raised external ion concentration rates of uptake increase and less specific activity is required for obtaining grain yields reasonable for grain counting (c) and microdensitometer readings (d). The densitometer measurements require somewhat higher grain yields than grain counting and hence, under the conditions of these experiments, a higher specific activity in the external solution is needed.

The results are shown on the graphs of Fig. 3.4 plotted as units $F_d \times \mu mo$-le and $F_c \times \mu mole$ versus external SO_4^{--} concentration. Since exposure times of the 3 experiments shown do not vary considerably, comparisons of the results obtained may be made although the proportionality factors have not been eliminated. For an example the results of densitometer readings from experiments (a) and (c) at the external SO_4^{--} concentration

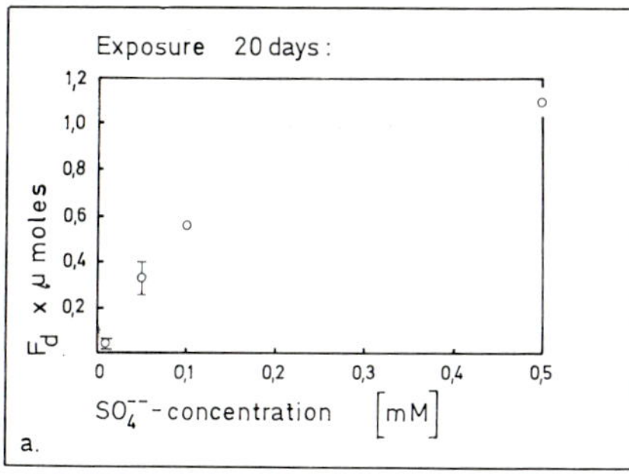

SO_4^{--}-conc. [mM]	Spec. Act. [mCi/mM]	Measure-ments
0.01	5900	d
0.05	1100	d
0.10	400	d
0.50	110	d

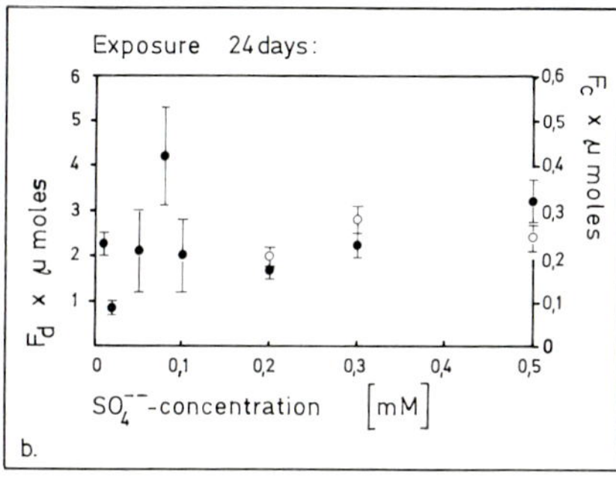

SO_4^{--}-conc. [mM]	Spec. Act. [mCi/mM]	Measure-ments
0.01	17	c
0.02	10	c
0.05	7	c
0.08	5	c
0.10	15	c
0.20	55	c,d
0.30	40	c,d
0.50	38	c,d

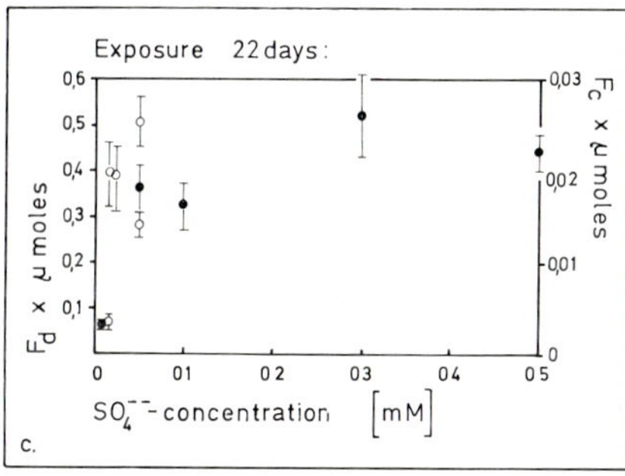

SO_4^{--}-conc. [mM]	Spec. Act. [mCi/mM]	Measure-ments
0.01	675	c
0.01	940	d
0.02	211	d
0.02	305	d
0.05	97	c
0.05	163	d
0.05	350	d
0.10	80	c
0.30	33	c
0.50	37	c

Fig. 3.4. Results of grain counts (●, index "c") and micro-densitometer measurements (○, index "d") of autoradiograms of *Nymphaea* hydropotes after 3 hrs of $^{35}SO_4^{--}$ uptake followed by 2 × 15 min of free space exchange in ice-cooled unlabeled uptake solutions. SO_4^{--}-concentrations and specific activities were varied as indicated. Thin tissue technique of autoradiography as described in chapter 3.4.2.6.1

of 0.05 mM may be compared. The results at specific activities varied between 163 and 1100 mCi/mM agree within reasonable limits:

experiment (a): $F_d \times \mu\text{mole} = 0.32 \pm 0.07$ at 1100 mCi/mM,
experiment (c): $F_d \times \mu\text{mole} = 0.51 \pm 0.05$ at 350 mCi/mM,
$F_d \times \mu\text{mole} = 0.28 \pm 0.03$ at 163 mCi/mM.

At 0.02 mM SO_4^{--} in experiment (c) the points obtained with the two different specific activities are very close to each other. This shows that with the data analysis provided by Eq. (3.1), specific activity in the uptake solution does not effect the final result. In the above tabulation variation among the results is most unlikely to be due to variation of specific activities, as the results for the most different specific activities are in best agreement (line 1 and 3).

Larger variations of the data are found in other instances, especially when the grain counts of experiments (b) and (c) are compared. Deviations between different experiments may be explained by the usual variations of tissue batches. In addition, the small differences in exposure time may add to slight differences of conditions at photographic development (time, age of developer, temperature) and thus affect the proportionality factors. In order to eliminate these variations the results of the 3 experiments were normalised by setting the results obtained for 0.1 mM external SO_4^{--} concentration at a given point and recalculating the rest of the data so that the ratios of results remained the same as in the original measurements (Fig. 3.5). In this way it can be shown that in the concentration range 0.01—0.5 mM the SO_4^{--} uptake into the hydropotes follows saturation kinetics. The hyperbolic curves drawn in Fig. 3.5 were calculated from the normalised measurements with the least squares method by using a linear transformation of the Michaelis-Menten equation (LÜTTGE et al., 1971).

Microautoradiographs at higher external concentrations were difficult to obtain, because with raised ion concentrations it becomes increasingly difficult to maintain a high specific activity in the external solution. A way out of this problem is to extend exposure time. After SO_4^{--} uptake by the

Table 3.2. Solutions for studying SO_4^{--} uptake into hydropote glands at high external SO_4^{--} concentration. With 20 days exposure some grain counting was possible, for micro-densitometry 150 days of exposure were required

External SO_4^{--} concentration [mM]	Specific activity (μCi/μM)
1.0	48
3.0	16
5.0	9
10.0	4

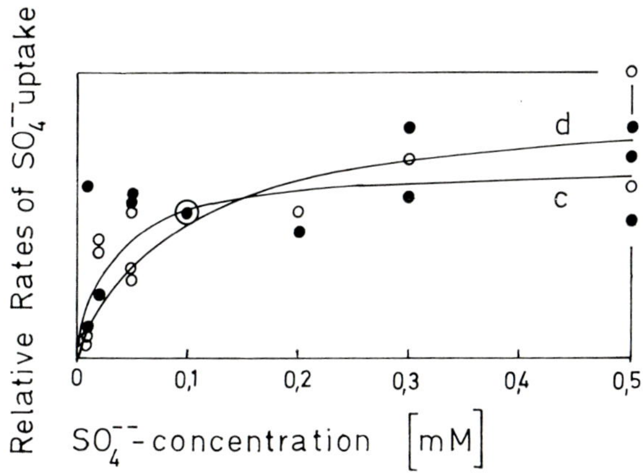

Fig. 3.5. Data of Figs. 3.4. a, b, c normalized at 0.1 mM SO_4^{--}-concentration (⊙) as described in the text. The curves for grain counts (c, ●) and densitometer measurements (d, ○) were calculated by the least square method using the following linear transformation of the Michaelis-Menten-equation (DOWD and RIGGS, 1965):

$$[S]/v = K_m/V_M + 1/V_M [S],$$

where [S] = sulfate conc.; V_M = maximal rate and v = actual rate of labeling; K_m = Michaelis constant. (LÜTTGE et al., 1971)

hydropote cells for 3 hrs from the solutions listed in Table 3.2 with 20 days of exposure only very low grain densities were obtained on the autoradiographs, but some grain counts could be made. With 150 days of exposure (~ 1.7 half-lives) reasonable grain yields for microdensitometry were obtained.

This detailed study of a specific problem provides rough guidelines of how various factors may be altered for a quantitative approach in microautoradiography. However, for each particular problem experience has to be newly gained.

3.3.2. The Usefulness of Static Autoradiographic Images in the Investigation of Dynamic Transport Mechanisms

The problem of interpreting static autoradiographic images in relation to dynamic events such as transport mechanisms is best illustrated by an analogy frequently used by Professor CRAFTS. It is the highway that serves as the route for traffic, yet a static image will show many more vehicles per unit area on a parking lot than on the highway itself. However, many other techniques also only provide data for the state at the sampling time, and the design of the experiments determines whether kinetics of a process can be examined using a particular technique.

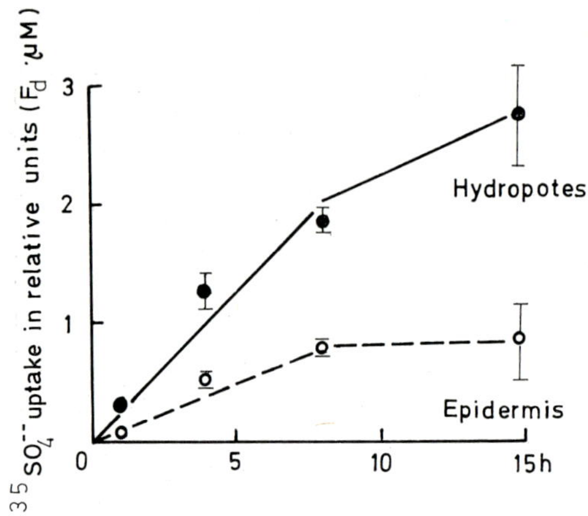

Fig. 3.6. Micro-autoradiographic determination of the time course of $^{35}SO_4^{--}$ uptake into hydropotes and epidermal cells of *Nymphaea* leaves. Labeled uptake solution: 0.05 mM $K_2{}^{35}SO_4$, spec. activity 400 mCi/mM; exposure time 45 days. Other details as in Figs. 3.4 and 3.5. (From LÜTTGE et al., 1971)

Taking tissue samples for preparation of microautoradiographs after varied times of application of a labeled ion solution will permit the time course of uptake of the ion into different parts of the tissue or into particular cells to be followed. The example depicted in Fig. 3.6 shows the time course of $^{35}SO_4^{--}$ uptake from a 0.05 mM SO_4^{--} solution into the hydropotes and epidermal cells of the lower leaf epidermis of *Nymphaea*. The finding that this time course is linear up to 8 hrs was important for the experiments shown in Fig. 3.4 and 3.5. The uptake time in the experiments of these figures was within the period of linearity (3 hrs !) and hence the units on the ordinates of Fig. 3.4 and 3.5 indeed represent relative rates of uptake at the various external concentrations. Thus taken together, the experiments of Fig. 3.4—3.6 represent a kinetic analysis of SO_4^{--} uptake into the hydropote gland cells. (A similar analysis has been made for epidermal cells, but for the purpose of illustrating the technical approach it does not seem necessary to include these data as well.) Since epidermal and hydropote cells of the *Nymphaea* leaf epidermis cannot be separated mechanically, histological techniques such as microautoradiography, and perhaps also electron probe analysis, are the only methods available to investigate the problem.

When ion transport through larger organs of plants (e.g. roots, stems etc.) is to be studied by microautoradiography sampling at various times after application of a labeled external solution may provide information on

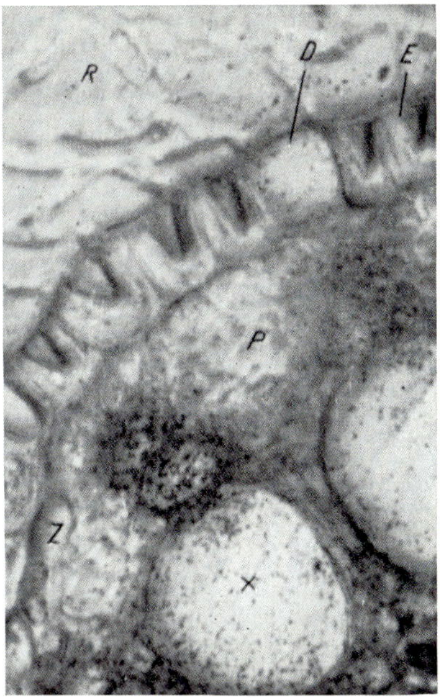

Fig. 3.7. Distribution of $^{35}SO_4^{--}$ in a cross section of an *Iris* root at a distance (ca. 5 cm) from the application of a 20 mM $K_2\,^{35}SO_4$ solution (spec. activity 5 mCi/mM). Precipitation technique of autoradiography. Note the high grain densities in the film above the xylem vessels indicating the labeling of the transpiration stream. Silver grains also appear above the passage cells (D) of the tertiary endodermis (E) and just adjacent to the passage cells in the cortex (R). Z=pericycle, P=phloem, X = xylem. Magnification 400 ×. (From ZIEGLER et al., 1963)

the routes of transport. In a transient period after application of the label before dynamic equilibria are established, the label will first appear in the routes of translocation, i.e. the "highways" will be labeled before the vehicles accumulate on the parking lots at the sides or at the destinations. Similarly, when labeled ions are transported along the length of an organ such as a root or a stem, and when a labeled ion solution is applied for a given time at a certain point, microautoradiographs taken at various distances from this point will allow conclusions regarding the route of translocation. Label will appear first in the tissues representing these routes and will only later build up at possible locations of accumulation. In experiments with *Iris* roots for example, microautoradiographs of cross sections were prepared at various distances from the point of application of a $^{35}SO_4^{--}$-solution. At a distance where no label was found in the cortex in general, xylem vessels where highly labeled, and there was some label in the passage cells of the tertiary endodermis and in a very restricted part of the cortex just adjacent to the passage cells (ZIEGLER et al., 1963; Fig. 3.7). This is highly suggestive for the passage cells providing a route for ion transport from the xylem vessels across the tertiary endodermis into the cortex.

Autoradiographs prepared in combination with pulse-chase experiments will also give insights into dynamic processes, and something is said on this in Chapter 5.4.3 and in Chapter 6 (6.6 and 6.16.3).

When they can be successfully used to describe different physiological states of tissues and the transient stages between those states, histological techniques may elucidate very dynamic aspects of a problem. The demonstration of the uptake of potassium by stomatal guard cells while these cells increase their turgor for stomatal opening, and of the release of potassium from guard cells when stomata close, is an excellent illustration of this (SAWHNEY and ZELITCH, 1969; HUMBLE and RASCHKE, 1971). Another outstanding example is the demonstration of K^+ movements between the dorsal and ventral parts of the pulvini of *Albizzia* leaflets during nyctinastic movements (SATTER et al., 1970). In both cases the potassium translocations have been demonstrated by electron probe analysis, however, for microautoradiographic investigations of dynamic events similar approaches would be appropriate.

In conclusion, microautoradiography is not inherently limited to investigations of static distribution of label in tissues and cells. The application will determine, whether the technique may be successfully applied in the investigation of dynamic events.

3.4. *Preparations for Microautoradiography of Water Soluble Inorganic Ions*

The preparative techniques for preventing leaching and artificial dislocation of label of water soluble inorganic and organic substances in microautoradiographic investigations and in electron probe analysis are similar. Hence some overlapping in the descriptions of technical approaches given in Chapter 4 and 7, and in this chapter are unavoidable. However, besides these similarities there are also important differences in the technical requirements. In particular: (i) while tissue sections for autoradiography have to be coated by a photographic emulsion, for electron probe analysis they have not; (ii) inorganic ions which are usually much more polar than, for example, products of ^{14}C-photosynthesis, may well penetrate polymers used for embedding the specimen for autoradiography of labeled organic compounds. Thus it appeared appropriate to give in this chapter a synoptic view of the techniques available for preparation in microautoradiography of inorganic ions and to make reference to chapters 4 and 7 where ever suitable.

3.4.1. The Stripping Film

The techniques described below have been designed to prepare tissue sections on microscopic glass slides ready to be coated with stripping film. The techniques do not exclude coating with liquid emulsions in principle, in fact preparation techniques required in experiments with liquid emulsions

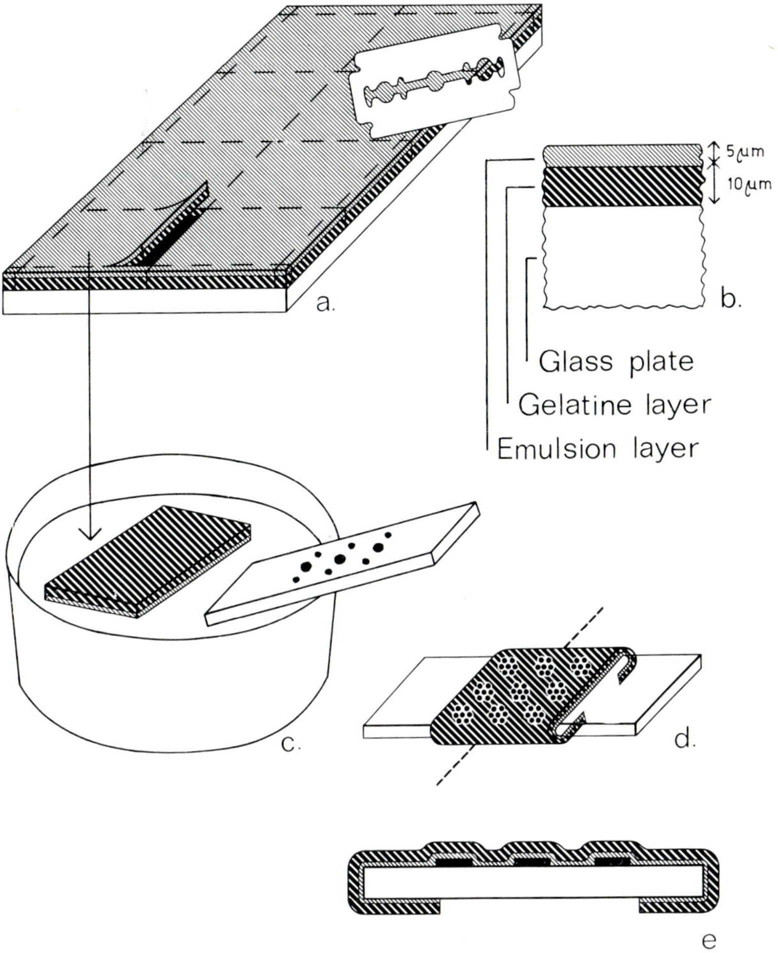

Fig. 3.8. Illustration of the stripping film technique. (a) Cutting the film on the glass plates ($4^3/_4 \times 6^1/_2$ inches) into suitable pieces. (b) Cross section of film and glass plate. (c) Floating a piece of film on the water bath, and coating the slide bearing the sections. (d) Coated slide. (e) Cross section of the coated slide

may be very similar. However, no experience with liquid emulsions has been gained in the author's laboratory. For a proper use of the stripping film a brief description may be helpful. (For details see PELC, 1956.)

Stripping film is provided by Kodak and the AR.10 fine grain quality will be appropriate for microautoradiography of most β-emitters such as those listed above (Table 3.1). The film is obtained on glass plates $4^3/_4 \times 6^1/_2$

inches in size. To facilitate handling the emulsion layer itself (ca. 5 μm thick) is carried on a 10 μm thick gelatin layer (Fig. 3.8a and b). The size of the developed silver grains is about 1 μm. The resolution obtainable under favourable conditions with this film is given by Kodak as 2.5 μm. Another stripping film (AR. 50) with a coarse-grain emulsion having an approximately 10× higher sensitivity has been available for some time. (Emulsion layer 12 μm thick, gelatin layer 10 μm thick, resolution under favourable conditions 12 μm; film not suitable for grain counting.) According to recent information from Kodak the production of the AR. 50 film has been discontinued.

For coating sections on a microscopic glass slide with the stripping film, the film on the glass plate is usually divided into 8 parts of approximately equal size by moving the tip of a razor blade across the film perpendicularly to the glass plate (Fig. 3.8a). In a similar way cuts are made along the edges of the glass plate (Fig. 3.8a). The pieces of film can now be removed from the glass plate using the edge of the razor blade, and they are then floated on a water surface in a glass dish (Fig. 3.8a and c). The part of the film that is going to cover the sections should not be touched with the fingers or any other object. This prevents mechanical damage which may produce a latent image.

Floating on the water surface the gelatin of the film will absorb water, swelling to a larger size within approximately 1 min. For optimal swelling a water temperature of 24 °C is suggested. During later photographic procedures (developing, rinsing, fixing and washing) performed at 18—20 °C less swelling will occur, and this will help to prevent loss of the film from the slides.

Microscopic glass slides are coated with the swollen film by gently moving the slides into the water underneath the film and "folding" the film around them (Fig. 3.8c). Intimate contact of the sections with the emulsion layer is required. Therefore in the example illustrated in Fig. 3.8 the film must be floated with the emulsion layer facing down towards the water surface. This generally applies to the autoradiography of water insoluble compounds in which sections mounted on the slide are coated with the wet film in the water bath. It also applies to some of the preparations in autoradiography of water soluble substances. However, in many other procedures (Fig. 3.9) the sections are not directly coated in the way illustrated in Fig. 3.8. Here care must be taken that the film is always floated such that the sections are finally in direct contact with the emulsion layer. Having chosen the preparative technique one must carefully observe that the film is floated in the correct way.

3.4.2. Specimen Preparation

3.4.2.1. General Remarks

The first step after loading a tissue with labeled ions will usually be to freeze the tissue as rapidly as possible to avoid artificial redistribution of the label (for exceptions see Chapter 3.4.2.6). Subsequent to freezing various techniques may be applied to obtain tissue sections, i.e. paraffin infiltration after freeze substitution, after freeze drying or after a precipitation, and sectioning in the cryostat (Fig. 3.9; Chapter 7.4.1; cf. also LÜTTGE and WEIGL, 1965).

The individual procedures used to obtain the sections are essentially identical to those in electron probe analysis, and the reader is referred to Chapter 7.4.1, pages 203—211. However, the section mounting techniques are different in microautoradiography and electron probe analysis, so that specific descriptions are required. Fig. 3.9 lists the procedures from the frozen plant material to the autoradiograph ready for microscopic inspection, which have been used.

In all circumstances, the microscopic glass slides used to bear the sections and film must be very clean. Cleanliness is the first requirement in microautoradiography, because dirt and dust particles left on the slides will cause artifacts in microphotography, grain-counting and microdensitometry of microautoradiographs. Many cleaning methods are available but those suggested by the Kodak instruction sheet contained in each package of stripping film are adequate (see also 5.6.1). To provide good wet-adhesion of the film and thus to prevent loss of the film at photographic processing the clean slides should be coated with gelatin (solutions of 0.1—1%) or with a gelatin-chrome alum solution as suggested by Kodak (0.5 gr. chrome alum + 5 gr. of gelatin in 1000 ml). (Cf. also 5.6.1.2 and 5.6.1.3 for the use of adhesives.)

3.4.2.2. Preparation I

Preparation I of Fig. 3.9 is the most generally applicable one. Paraffin blocks of the tissue are obtained after freeze substitution with ethanol, ether or acetone, and after vacuum infiltration of the freeze dried material respectively (for details see Chapter 7, Sections 7.4.1.3 and 7.4.1.4). Cryostat sections are prepared as described in Chapter 7.4.1.1.

Sections obtained by either of these methods are placed on a microscopic glass slide, which has been coated with stripping film and then thoroughly dried. All operations are done in the darkroom under safelights. In the case of cryostat sections these operations must be done in the cold i.e.

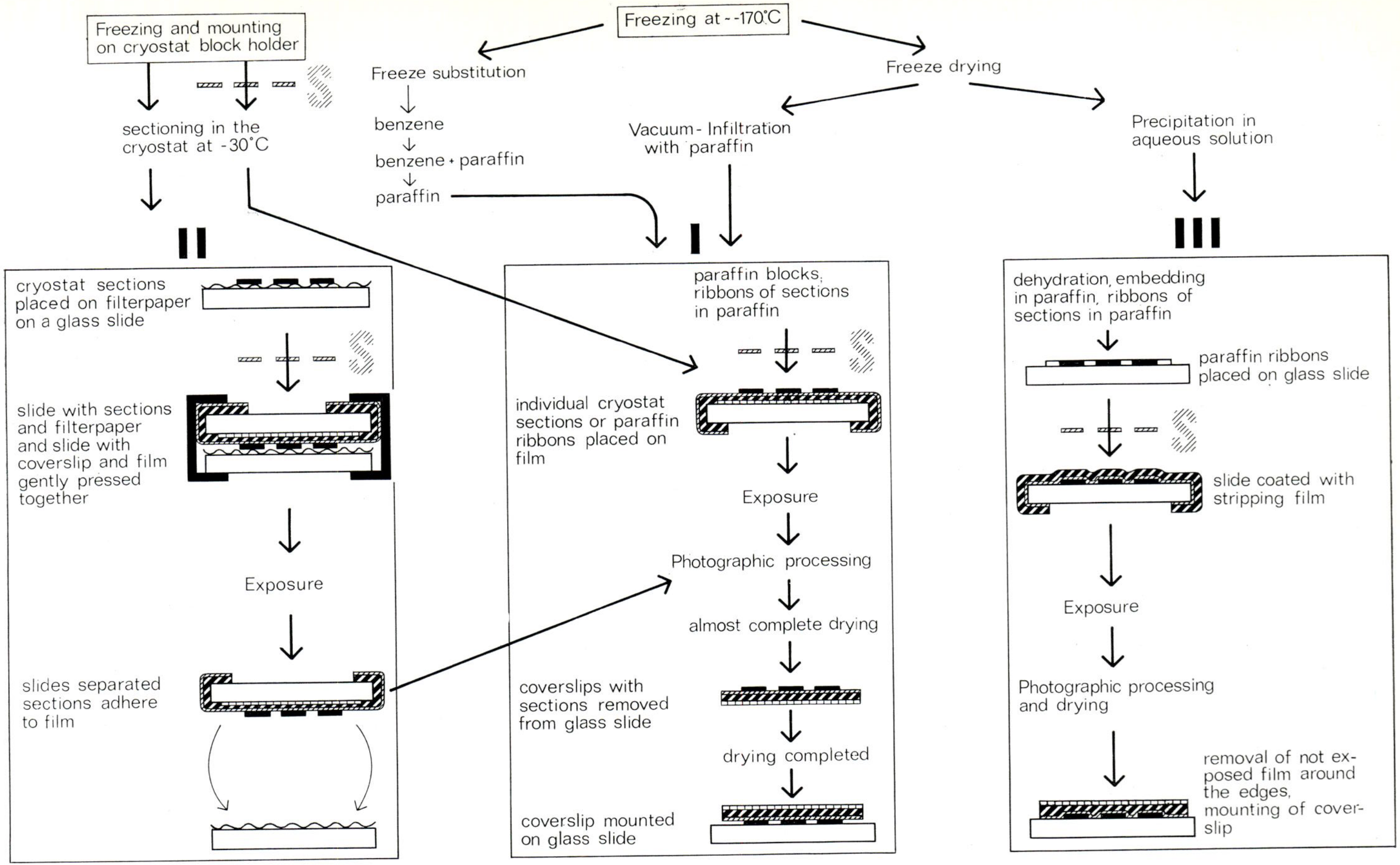

Fig. 3.9. Chart summarizing various preparation techniques as explained in the text. Hatching of the various layers of sections and film is the same as used in Figs. 3.8. and 3.10. S=safelight.

within the cryostat chamber. The addition of a drop of cold hexane + liquid paraffin helps to establish good contact between the sections and the film (TAUGNER et al., 1958a and b). Paraffin sections may be placed on the film at room temperature. A droplet of ether will provide good contact.

The slides used are coated with the stripping film in such a way, that the emulsion layer faces upwards in direct contact with the sections placed on top. The suggestion of BRANTON and JACOBSON (1962) to put a large microscopic coverslip (vertically hatched in Fig. 3.9) on the glass slide before coating, so that the coverslip eventually is placed underneath the film, is strongly recommended. This allows the preparation to be turned upside down at the end of the procedure (see below) so that the autoradiograph can be viewed in the microscope through the thin coverslip and the thin gelatin layer of the film. Viewing through the thick microscopic glass slide prevents the use of high magnifications, and viewing through the tissue section itself is difficult because cell walls and cell contents obscure the silver grains in the emulsion layer of the film.

During exposure moisture must be kept away from the preparations. Incidentally, this also applies for autoradiography of non-water soluble compounds, because a moist atmosphere will cause high background and blackening of the film after photographic development. Microorganisms may also grow on the gelatin of the film. For these reasons it is advisable to expose in a deep freeze at about —20 to —30 °C. Slides prepared in the cryostat are quickly transferred from the cryostat chamber to the deep freeze using a tight box.

It is suggested to press pairs of microautoradiographic sandwiches together back to back and to place them in a rack as shown in Fig. 3.10f. (Suitable racks are best made from "plexiglass".) This can be done either at the beginning or at the end of exposure. It will facilitate handling in the subsequent steps of photographic processing, and will also help to prevent the film from floating away from the slides. (Similar provisions should be made in preparations II and III.)

After exposure, racks with the slides are taken from the deep freeze and allowed to warm up to room temperature within a dessicator. Paraffin sections are then freed of paraffin rinsing two times (approximately 10 min each) in two changes of xylene (check if paraffin is completely removed!). The xylene is removed by two changes (2×5 min) of absolute ethanol (or any other low carbon-chain absolute alcohol, e.g. absolute isopropanol). The alcohol can then be evaporated directly from the slides, or alternatively the slide can be transferred into water through alcohol changes in a number of steps (80%-60%-40% alcohol-water).

Cryostat slides and paraffin free slides are dipped into a 0.5—1% gelatin solution and then dried. Gelatin coating has been found to minimise loss of sections from the film during photographic processing. Photographic processing is identical for the various preparations and will be dealt with in a separate chapter (3.4.3).

Before the developed, fixed and washed autoradiographs are completely dry the coverslips bearing the film and the sections are removed from the supporting glass slides by cutting along the edges with a razor blade. It will be difficult to do this, when the film is completely dry, but the film must be dry enough to prevent movements of film and sections of the autoradiographs. Safety eye glasses should be worn because the coverslips may break during this procedure! The film is then dried completely. Now the coverslips may be turned upside down, mounted appropriately on a clean glass slide using a suitable mounting medium (caedax, canada balsam, epoxy resin or the like), and the autoradiographs are ready for microscopic inspection.

3.4.2.3. Preparation II

Preparation II is a variation suggested by (KOLLMANN, 1965, 1967). The main advantage of this procedure is that it is possible to perform the sectioning in the cryostat under full normal illumination. Sections are placed on top of a piece of filter paper carried on a microscopic glass slide. Safelights are only required after collecting the sections when the slides with the filter paper and the sections are pressed against slides coated with the stripping film. The slides with the film are similar to those used in preparation I, but in addition it appears to be important that the film is covered by a thin layer of chrome alume-gelatine or egg-white-glycerin (KOLLMANN, 1965, 1967). The disadvantage of this is that it will reduce the resolution of the microautoradiographs because an extra layer is introduced between the emulsion and the sections. However, according to KOLLMANN this provision is necessary to insure that the sections stick to the slide with the film, when the two slides are separated at the end of exposure time (Fig. 3.9). The rest of the procedure in preparation II is identical to that of preparation I.

The variation II has been successfully used by KOLLMANN and his coworkers in microautoradiography of organic ^{14}C-compounds translocated in the phloem of wooden stem tissues. There is no doubt it will be similarly useful in the autoradiography of inorganic ions. However, this technique is probably restricted to work with sections of rather hard materials and may turn out difficult to perform with sections of soft tissues.

3.4.2.4. Preparation III

Preparation III is essentially that used in microautoradiography of nonwater soluble compounds as shown in Fig. 3.8. After freeze drying of the frozen plant material, precipitation of the water soluble ions in a non-soluble form is achieved by immersion of the tissue in an aqueous solution of the precipitating reagent as described in Chapter 7.4.1.2. The tissue is dehydrated and embedded in paraffin in the usual way, going through changes of alcohol in water with increasing alcohol concentration, benzene and benzene + paraffin into pure paraffin (Chapter 7). Paraffin sections are mounted on the glass slides coated with chrome alum-gelatin and the stripping film is applied as depicted above (Fig. 3.8). After exposure and photographic processing the non-exposed film around the edges may be removed. The autoradiographs can be inspected through the protective gelatin layer of the film without adding a coverslip, or alternatively a coverslip may be mounted.

3.4.2.5. Embedding in Polymerising Resins

A variation of procedure involving methacrylate embedding as used in electron-microscopy (PALADE, 1952) is described by GIELINK et al. (1966) as follows below. This technique was used in microautoradiography of $^{45}Ca^{++}$ distribution in oat plant tissues.

The frozen tissue is freeze substituted in water free acetone at —80 °C. In the initial 24 hrs the acetone contains 1% OsO_4 as a fixing reagent improving preservation of cytological structures. Further changes of acetone renewed daily are free of OsO_4. At the end of freeze substitution (approximately 6 days) the tissue is transferred to a monomer methacrylate mixture (methacrylate : butylmethacrylate 1 : 4) at —50 °C, which is changed 3 times. Towards the end of infiltration with that mixture the tissue is allowed to warm slowly to room temperature and is introduced into gelatin capsules filled with the polymerising methacrylate mixture.

Sections (4—8 μm thick) are placed on chrome alum-gelatin covered glass slides, stretched on a drop of water and allowed to dry. Another gelatin coating is then added (0.5% gelatin solution). The stripping film is applied wet as in microautoradiography of non-water soluble substances (Fig. 3.8, Preparation III in Fig. 3.9).

The crucial question regarding the application of this technique in microautoradiography of inorganic ions is whether methacrylate embedding after polymerisation entirely prevents leaching and redistribution of the labeled ions when the tissue sections are allowed to contact water. It must be assumed that methacrylate infiltration thoroughly protects all parts of the sections from any contact with water. No radioactivity was detected by liquid scintillation counting of the water used for swelling the film and for stretching the sections. The organic solvents used in the preparation were also tested for radioactivity, but only a minimal loss of label from the tissue was detected.

Embedding in polymerising Durcupan ACM resin has been successfully used by Fritz and Eschrich (1970) in microautoradiographic studies with ^{14}C-labeled organic compounds (see Chapter 4). However, inorganic ions are much more polar, and therefore there are still some doubts, whether these approaches are also suitable in the work with ions. Polar particles may find some routes leading through the resins. This suspicion is amplified by the experience that ionic contrasting reagents (e.g. lead-citrate and uranyl-acetate, Reynolds, 1963; Peters et al., 1971) are readily introduced into resin embedded sections, a staining technique widely used in electron-microscopy. Loss of activity from the sections might have been overlooked in the experiments of Glielink et al. (1966) because of the relatively large volumes of water brought in contact with the thin and tiny sections. In an electron probe analysis study of the distribution of water soluble ions in maize roots epoxy resin embedding has proved as a very suitable approach to retain the ions at the original sites in the tissue. However, in this case water has been kept away from the specimens throughout the entire preparation, i.e. not only during embedding itself but also at sectioning and at mounting of the sections (Läuchli et al., 1970; see Chapter 7.4.1.4). It is felt that more trial experiments in microautoradiography of inorganic ions using the methacrylate embedding procedure should be run before this technique is more generally adopted.

3.4.2.6. Preparation with no Sectioning

Although highly depending on the tissue and limited to a particular sort of investigations two techniques avoiding any sectioning of the plant material after labeling might be useful for the reader to approach his problems and should be discussed.

3.4.2.6.1. *The Thin Tissue-Method.* The use of thin tissues which can be directly coated with the stripping film avoids any sectioning during preparation of the autoradiographs. This saves time and excludes artifacts due to sectioning and related steps in preparation. Tissues of a thickness of only one cell layer, e.g. moss leaves, epidermal strips of onion scales (*Allium cepa*) and of *Chenopodium* leaves (see Fig. 1, introductary chapter) have been used in the author's laboratory. In the experiments on $^{35}SO_4^{--}$-uptake into hydropotes and epidermal cells of *Nymphaea* leaf epidermis discussed above (Chapter 3.3), prior to isotope uptake thin sections (3—5 mm in diameter) were cut parallel to the surface of the leaves using a sharp razor blade. Epidermal cells and cells of the hydropote gland complex were the only remaining intact cells in these thin sections. The material thus obtained was directly covered with stripping film at the end of the labeling experiments.

The method should be also ideal for autoradiography of algal chains or smears of unicellular algae.

In this procedure the wet stripping film swollen on water is applied to the wet tissue after which the preparation is rapidly frozen in the deep freeze. In the time required to assemble the autoradiographic sandwich as described below, leaching of label from the tissue is possible. Therefore the method is only applicable in experiments designed so that this leaching is minimal and does not affect the results which one wants to obtain. In fact, this applies to many ion uptake experiments. Usually, after ion accumulation in a labeled solution, plant tissues are washed in changes of ice-cooled unlabeled solutions to exchange labeled ions from the free space, so that only label taken up beyond the permeability barrier of the cells is detected. Time of free space washing is frequently in the range of 10—30 min. The time required for arranging the pieces of thin tissue and the stripping film for exposure in the deep freeze is in the order of 1 min or less. This time is negligible compared with the duration of free space washing. Hence, the technique described here should be quite appropriate in many experiments involving exchange of free space.

The details of the procedure of coating the tissue with the stripping film are shown in Fig. 3.10. The washed tissue is placed on microscopic glass slides which have been covered with a gelatin layer. The slides are transferred to the dark room (Fig. 3.10a). In a water bath at 24 °C stripping film is stripped over a blank microscopic slide with the sensitive photo-emulsion facing upwards (Fig. 3.10b). The two slides, one bearing the tissue and the other bearing the film, are gently pressed together to give good contact between the film and the tissue (Fig. 3.10c). The edges of the wet film are then moved around from the blank slide to the slide bearing the tissue as shown in Fig. 3.10d, and the blank slide is removed (Fig. 3.10e). Two of the slides coated with the film now are pressed together back to back (Fig. 3.10f). The "sandwiches" obtained are placed in a rack and transferred to the deep freeze for exposure.

To improve resolution when the distribution of labeled ions between the cytoplasm and the vacuoles is to be determined, the thin tissues may be centrifuged during free space washing prior to autoradiography. In this way the cytoplasm plus the enclosed organelles is deposited in one part of the cells, and radioactivity in the cytoplasm and vacuoles is easily resolved (LÄUCHLI and LÜTTGE, 1968; Fig. 3.11). In onion scale epidermis cells cytoplasmic streaming within 10 min after centrifugation redistributes the cytoplasm in the normal way, so that the centrifuged cells look identical to cells of controls. Neutral red staining of centrifuged onion and moss cells has also confirmed that the integrity of the vacuoles is maintained and that no irreversible damage is done to the cells by centrifugation. This shows that with this technique the radioactivity in the cytoplasm and in the vacu-

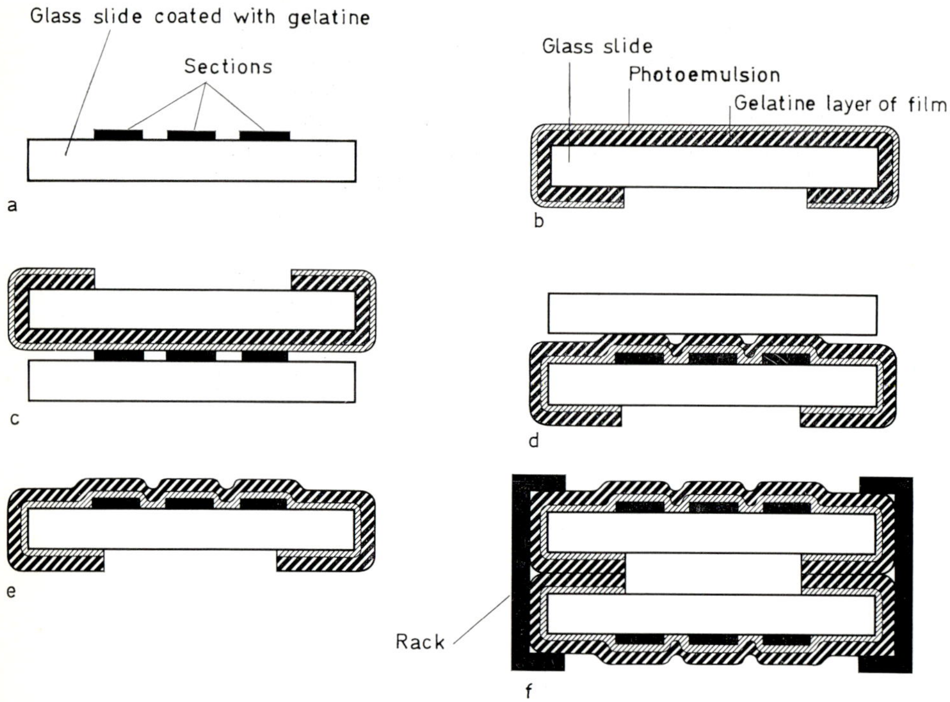

Fig. 3.10. Schematic representation of the "thin tissue technique" of micro-autoradiography. (a) Sections are placed on gelatine-coated slides. (b)—(f) Dark room procedures: (b) a blank glass slide is coated with stripping film; (c) the slide with the sections and the slide with the film are brought in contact; (d) the edges of the film are stripped around the slide with the sections; (e) after removal of the blank slide, the sections are properly coated with the film; and (f) micro-autoradiographic sandwich is stored in a rack for exposure and for later photographic development and fixation (After LÜTTGE et al. 1971)

oles is correctly measured. By contrast, with techniques involving sectioning especially after freeze drying, there is much uncertainty whether the content of large sap vacuoles is preserved *in situ* during preparation for micro-autoradiography.

However, the problem of differential shrinkage of frozen sections, discussed by CLARKSON and SANDERSON (1970), is still pertinent. Frozen sections and tissue slices will loose water by sublimation during exposure in the deep freeze. By this freeze drying the various parts of a cell (i.e. cell wall, cytoplasm, vacuole) or of a tissue will shrink to different extents because they loose different amounts of water. Thus a section of an initially uniform thickness will dry to a state where thickness of cell walls will have changed slightly, cytoplasm will have shrunken considerably, and the sap

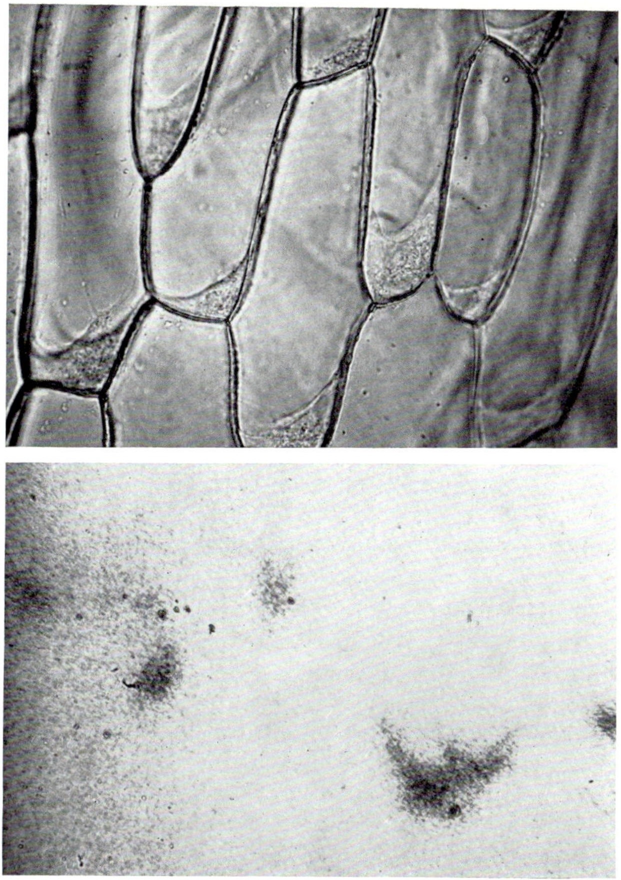

a

b

Fig. 3.11. (a) Centrifuged cells of a stripped adaxial onion (*Allium cepa*) scale epidermis. (b) Autoradiogram of a centrifuged stripped epidermis after labeling with $^{35}SO_4^{--}$. The tissue has been removed from the autoradiogram. "Thin tissue method" of autoradiography. Magnification 260 ×. (From Läuchli and Lüttge, 1968)

contained in the vacuoles and in dead tissue elements (e.g. xylem vessels) will have dried to a very thin layer (Clarkson and Sanderson, 1970). Since the resolution of autoradiographs is inversely related to the thickness of the sections, the contribution of radioactivity in the cell walls to the grain numbers detected above adjacent cytological structures will be considerably greater than the crossfire from radioactivity in the cytoplasm and in the vacuole. Artifacts in the grain distribution above various cytological structures may thus originate (Fig. 20 in Clarkson and Sanderson, 1970).

3.4.2.6.2. *The Sellotape Method.* The Sellotape method has been used in the author's laboratory to obtain autoradiographs of epidermal bladder and stalk cells of *Atriplex spongiosa* leaves (Osmond et al., 1969). The method is briefly mentioned here for the sake of curiosity, but it also may prove applicable in the autoradiography of other surface structures. After labeling with $K^{36}Cl$ the leaves were placed on a strip of Sellotape which was gently pushed against both surfaces. It was removed with a large number of adhering bladders and stalk cells. This was immediately stuck to a slide covered with stripping film and exposed in the deep freeze. The Sellotape was removed before photographic processing. The results suggested high amounts of label in the dense stalk cytoplasm and in the thin cytoplasmic layer along the walls of the bladder cells, and this was confirmed by autoradiographs obtained from paraffin sections after vacuum embedding (Fig. 3.12).

3.4.3. Photographic Processing and Staining

Suggestions for photographic processing are given on the information sheet contained in the film packages. Following these suggestions carefully one will obtain adequate results. For a more sophisticated discussion of the photographic process the reader is referred to monographs like those by BOYD (1955) and ROGERS (1967).

The exposed slides, treated to remove paraffin where necessary (see above), are usually developed for about 5 min at 18—20 °C. Kodak D 19b developer has been used in all of the author's experiments. Extension of development times up to 20—30 min is possible and can be used to increase speed of the film, but resolution will decrease at the same time. After development, slides are rinsed briefly with clean water (30 sec, at 18—20 °C) and then immersed in the fixation bath. After fixation slides are thoroughly washed in running tap water for at least 2—3 min (much longer washing may be required) and dried at room temperature in a dust free atmosphere.

If required, tissues should be stained *after* photographic processing. Staining the tissue prior to autoradiography will be usually impossible in experiments with water soluble compounds. Furthermore this is often inappropriate because it increases the danger of chemical fogging or chemical desensitisation respectively of the film. It is important that the staining technique chosen preferentially stains the tissue and that the dyes are absorbed by the gelatin of the film to a much lesser extent. For most plant tissues staining will be unnecessary. (For staining techniques see also Chapters 1.5.4 and 5.11.)

Chemical fogging, chemical desensitisation or latent image fading may not only be caused by stains introduced in the tissue but also by chemical agents inherent in the tissue. This may lead to the severest artifacts, and the appropriate controls should be run whenever a new type of plant material is used.

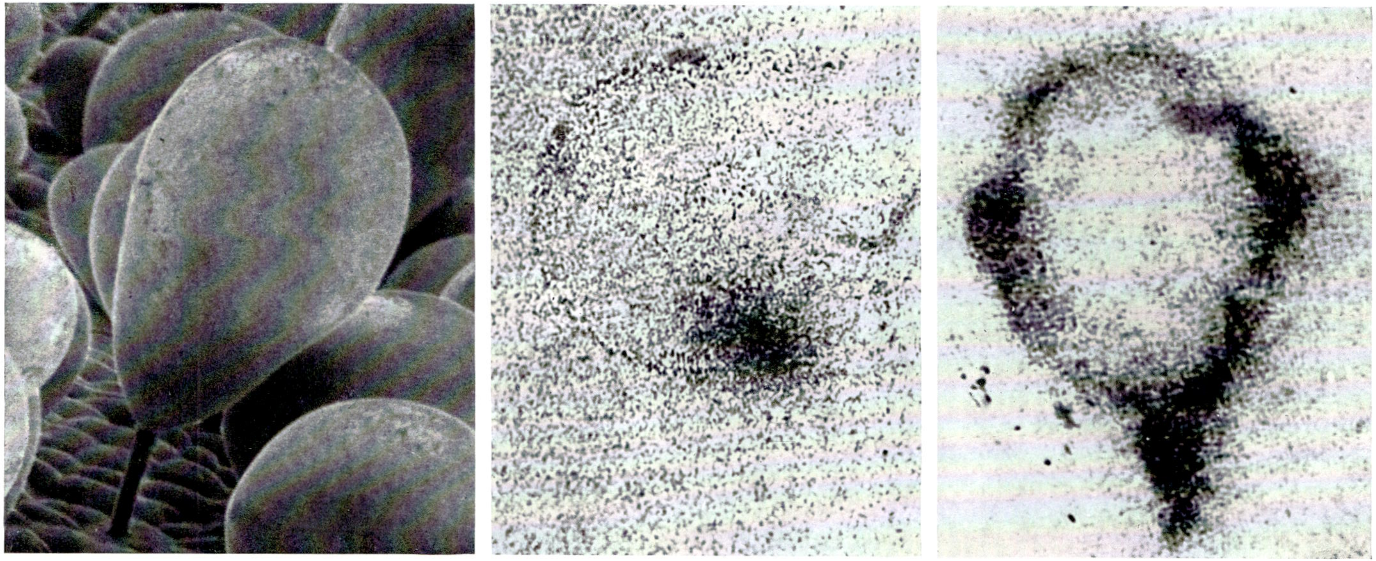

a b c

Fig. 3.12. (a) Scanning-electron microscope picture of *Atriplex spongiosa* epidermal bladder and stalk cells. (b) Autoradiogram of a bladder with its stalk cell, prepared using the Sellotape-method. Distribution of label in the bladder and stalk cells suggests high labeling of the cytoplasm. There was also label above background detected on the film outside the bladders, presumably due to leakage from the bladders during preparation. (c) Autoradiogram of a section through a stalk and a bladder prepared with the paraffin embedding technique (preparation I) confirming the results of (b). Magnifications approximately 600 ×. [(a) by courtesy of Dr. J. Throughton, Wellington, New Zealand. (b) and (c) from Osmond et al., 1969]

3.5. *The Documentation of the Results*

The techniques described in the preceding parts of this chapter provide steps, which are designed to prevent dislocation of sections on the film or loss of sections during photographic processing. Nevertheless, a certain proportion of sections is lost, but one will usually obtain enough sections in intimate contact with the film. This has to be checked carefully. To avoid these problems it has been suggested that every second section of a series be put aside for comparison with the autoradiogram (BOYD, 1955, TAUGNER et al., 1958a and b). However, it will often prove difficult to allocate an autoradiogram exactly to the part of the tissue section from which it arose.

On the other hand, if the distribution of silver grains on the autoradiogram allows very clear distinctions of various tissues or cell types, e.g. in microautoradiographs of stomata (Fig. 1, introductary chapter) or of hydropotes (Fig. 3.3), it may be quite useful to have the sections separated from their autoradiograms. In particular this will facilitate microdensitometry in the transmitted light, because no correction for light absorption by the tissue itself is required. Such a correction would turn out to be impossible in most cases (see Chapter 1).

If the sections are to be separated from the film at the end of photographic processing, it is often enough to avoid all the precautions to maintain contact suggested above. In our experiments using the thin tissue method described in Chapter 3.4.2.6.1 we removed the dried film with the autoradiograms from the tissue underneath it using razor blades, needles and fine foreceps. The isolated autoradiograms were put on a droplet of lens immersion oil on a slide, were covered with a coverslip and then photographed or subjected to microdensitometry.

Results are frequently documented by microphotography. It is often necessary to take two shots, focussing the tissue and the autoradiogram respectively. Microphotographs may be taken both in the transmitted light where the developed silver grains will appear as black dots or alternatively in incident light where the reflecting silver grains appear as bright dots, or by a combination of both (Fig. 3.13). Microphotography is naturally non-

---

Fig. 3.13. Autoradiogram of a *Nepenthes* pitcher-wall cell labeled with $^{36}Cl^-$. Paraffin infiltration technique (preparation I). (a) Micro-photography in the transmitted light: developed silver grains appear as black dots, the cytological structures are clearly visible. Label is largely found in the cytoplasm especially in some cytoplasm which has accumulated around the nucleus. (b) Micro-photography in the incident light. Silver grains appear as bright dots. Little is discernable of cytoplasmic structures. (c) Micro-photography with successive exposures in the transmitted and in the incident light respectively. The silver grains appear as bright dots *and* the cytoplasmic structures are visible. Magnification 1900 × [(a) original, (b) and (c) from LÜTTGE, 1966]

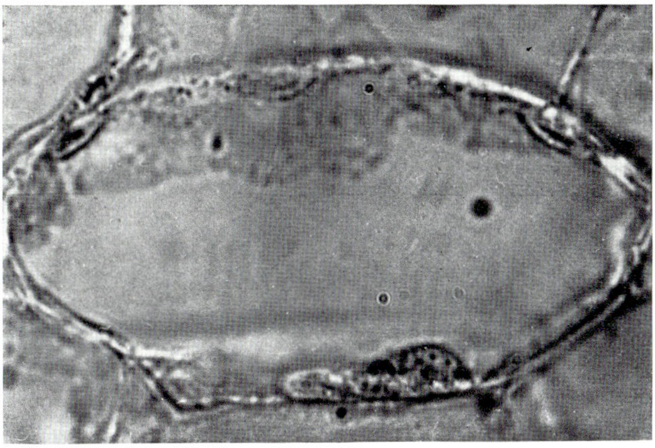

a

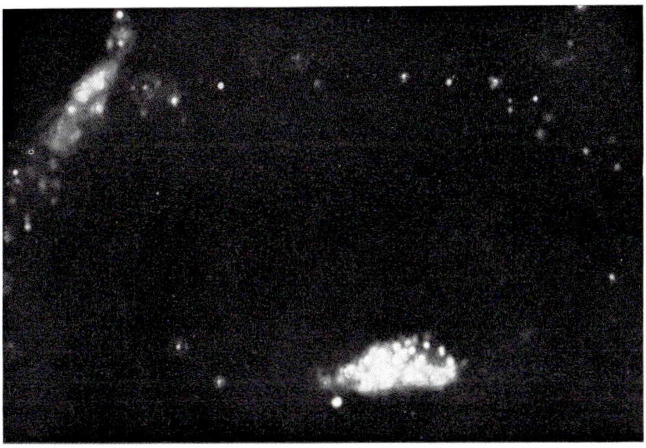

b

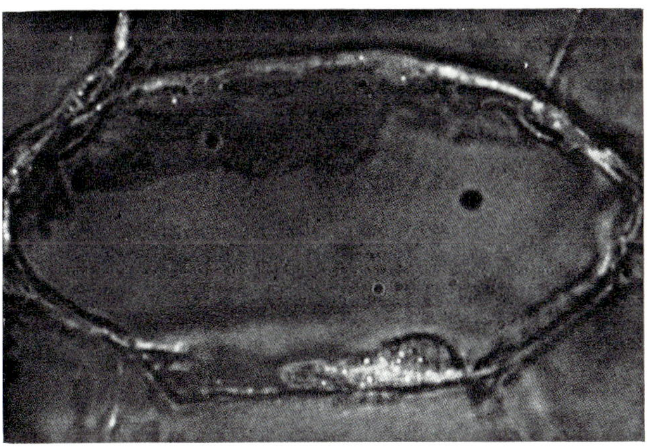

c

quantitative or at the most semi-quantitative as far as it allows rough quantitative evaluations by visual comparison.

Some ways for quantitative measurements are briefly mentioned above (Chapter 3.3.1). The photometric approach is considered in detail in Chapter 1. Quantitative measurements may be performed as point analysis (e.g. Fig. 3.4—3.6) or by scanning across a section (e.g. Fig. 3.14). Equilibrations by autoradiographing standards containing known amounts of radioisotopes (cf. Chapter 1) have been rarely achieved in work with plant material so that most quantitative measurements reported in the literature are expressed in relative units.

In the microautoradiographic experiments described in this chapter documentation of

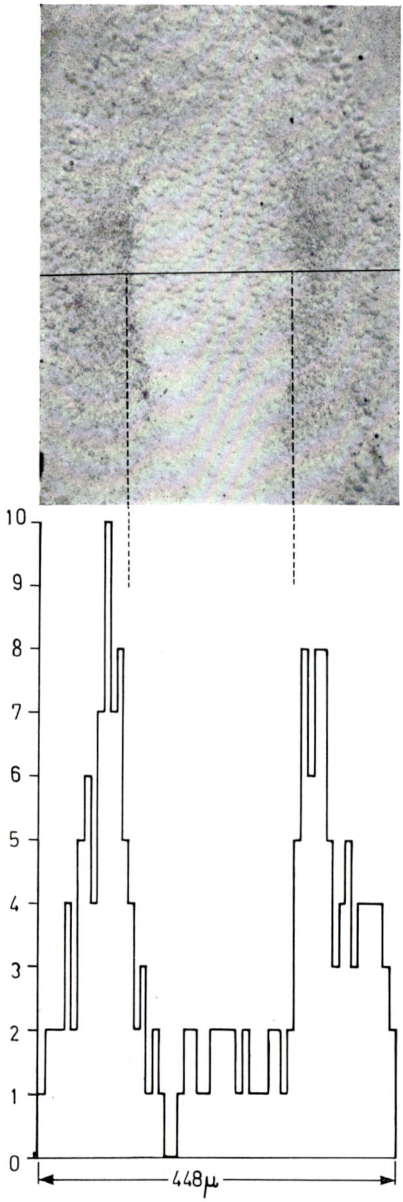

a

b

Fig. 3.14. (a) Micro-autoradiogram of a cross section through a young pea root labeled with $^{35}SO_4^{--}$. (SO_4^{--} uptake for 5 min from a 20 mM SO_4^{--}-solution, spec. activity 0.5 mCi/mM, no free space exchange, exposure 60 hrs; precipitation technique of autoradiography). (b) Grain counts along the solid line (scan) shown in (a). The labeled sulfate has penetrated the periblem and label has accumulated at the boundary between periblem and plerom. (From LÜTTGE and WEIGL, 1962) the results is based on interpretation of random grain images, e.g. photography and quantitative measurements of the random distribution of grains in the autoradiogram. The alternative, i.e. the interpretation of track images, has been rarely used in microautoradiography of inorganic ions in plant tissues and very little experience has accumulated (e.g. Fig. 3.15). The advantage of track autoradiography versus random grain autoradiography is discussed by BOYD (1955). Artifacts due to chemical fogging or high background caused by light or even misinterpretations of mechanical scratches are impossible, because tracks in the film with a width of one grain may be solely caused by radioactive disintegrations. A high resolution is obtained by tracing the tracks back to their origin (cf. Fig. 3.15). However, background tracks will always be present, and a careful evaluation of background and statistical analysis are essential.

Acknowledgements

The author's own experience with the use of autoradiographic techniques has been made possible by grants from the Deutsche Forschungsgemeinschaft. Dr. C. B. Osmond (Canberra) was kindly reading and correcting the manuscript.

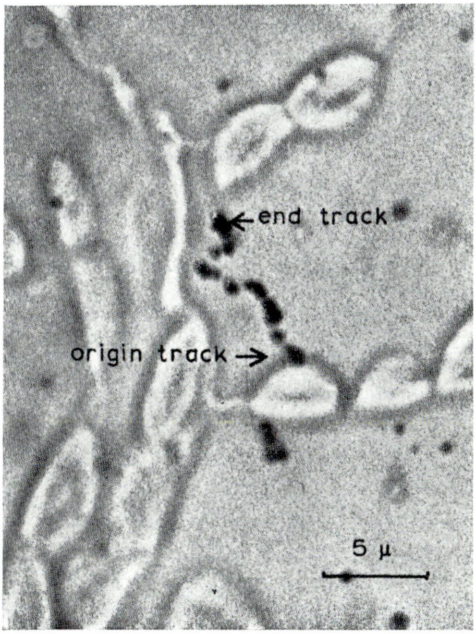

Fig. 3.15. Track image from an autoradiogram of an oat leaf cell labeled with $^{45}Ca^{++}$. The origin of the track is above a chloroplast suggesting $^{45}Ca^{++}$ uptake into the chloroplast. Magnification 2800×. Methacrylate embedding technique of autoradiography. Section thickness 6 µm. Ilford K 5 liquid emulsion. (With kind permission of Dr. A. RINGOET from RINGOET et al., 1971b; for further details consult this publication.)

References

BARBER, D. A., SANDERSON, J., RUSSELL, R. S.: Influence of micro-organisms on the distribution in roots of phosphate labeled with phosphorous-32. Nature (Lond.) **217**, 644 (1968).

BIDDULPH, S. F.: Visual indication of ^{35}S and ^{32}P translocation in the phloem. Amer. J. Bot. **43**, 143—148 (1956).

BIDDULPH, S. F.: A microautoradiographic study of Ca^{45} and S^{35} distribution in the intact bean root. Planta (Berl.) **74**, 350—367 (1967).

BOYD, G. A.: Autoradiography in biology and medicine. New York: Academic Press 1955.

BRANTON, D., JACOBSON, L.: Dry, high resolution autoradiography. Stain Technol. **37**, 239—242 (1962a).

BRANTON, D., JACOBSON, L.: Iron localisation in pea plants. Plant Physiol. **37**, 546—551 (1962b).

CLARKSON, D. T., SANDERSON, J.: The uptake of a polyvalent cation and its distribution in the root apices of *Allium cepa*: Tracer and autoradiographic studies. Planta (Berl.) **89**, 136—154 (1969).

CLARKSON, D. T., SANDERSON, J.: A critical review of certain autoradiographic techniques. Agricult. Res. Council. ARCRL **20**, 68—73 (1970).

CLARKSON, D. T., SANDERSON, J., RUSSELL, R. S.: Ion uptake and root age. Nature (Lond.) **220**, 805—806 (1968).

DOWD, J. E., RIGGS, D. S.: A comparison of the estimates of Michaelis-Menten kinetic constants from various linear transformations. J. biol. Chem. **240**, 863—869 (1965).

FRITZ, E., ESCHRICH, W.: ^{14}C-Mikroautoradiographie wasserlöslicher Substanzen im Phloem. Planta (Berl.) **92**, 267—281 (1970).

GAHAN, P. B., RAJAN, A. K.: Autoradiography of $^{35}SO_4$ in plant tissue. Exp. Cell Res. **38**, 204—207 (1965).

GIELINK, A. J., SAUER, G., RINGOET, A.: Histoautoradiographic localisation of calcium in oat plant tissues. Stain Technol. **41**, 281—286 (1966).

HUMBLE, G. D., RASCHKE, K,: Stomatal opening quantitatively related to potassium transport. Plant Physiol. **48**, 447—453 (1971).

JESCHKE, W. D.: Über die Verwendung von ^{86}Rb als Indikator für Kalium, Untersuchungen am lichtgeförderten ^{42}K/K- und ^{86}Rb/Rb-Influx bei *Elodea densa*. Z. Naturforsch. **25**b, 624—630 (1970).

KOLLMANN, R.: Zur Lokalisierung der funktionstüchtigen Siebzellen im sekundären Phloem von *Metasequoia glyptostroboides*. Planta (Berl.) **65**, 173—179 (1965).

KOLLMANN, R.: Autoradiographischer Nachweis der Assimilat-Transportbahn im sekundären Phloem von *Metasequoia glyptostroboides*. Z. Pflanzenphysiol. **56**, 401—409 (1967).

KRICHBAUM, R., LÜTTGE, U., WEIGL, J.: Mikroautoradiographische Untersuchung der Auswaschung des „anscheinend freien Raumes" von Maiswurzeln. Ber. dtsch. bot. Ges. **80**, 167—176 (1967).

LÄUCHLI, A., LÜTTGE, U.: Untersuchung der Kinetik der Ionenaufnahme in das Cytoplasma von *Mnium*-Blattzellen mit Hilfe der Mikroautoradiographie und der Röntgen-Mikrosonde. Planta (Berl.) **83**, 80—98 (1968).

LÄUCHLI, A., SPURR, A. R., WITTKOPP, R. W.: Electron probe analysis of freeze-substituted, epoxy resin embedded tissue for ion transport in plants. Planta (Berl.) **95**, 341—350 (1970).

LÜTTGE, U.: Über die Zusammensetzung des Nektars und den Mechanismus seiner Sekretion. III. Die Rolle der Rückresorption und der spezifischen Zuckersekretion. Planta (Berl.) **59**, 175—194 (1962).

LÜTTGE, U.: Mikroautoradiographische Untersuchungen über die Funktion der Hydropoten von *Nymphaea*. Protoplasma (Wien) **59**, 157—162 (1964a).

LÜTTGE, U.: Mikroautoradiographischer Nachweis der Aufnahme von $^{35}SO_4^{--}$ in die Wurzelhaare von Ahornkeimlingen. Naturwissenschaften **51**, 296—297 (1964b).

LÜTTGE, U.: Untersuchungen zur Physiologie der Carnivoren-Drüsen. II. Über die Resorption verschiedener Substanzen. Planta (Berl.) **66**, 331—344 (1965).

LÜTTGE, U.: Untersuchungen zur Physiologie der Carnivoren-Drüsen. V. Mikroautoradiographische Untersuchung der Chloridsekretion durch das Drüsengewebe von *Nepenthes*. Planta (Berl.) **68**, 269—285 (1966).

LÜTTGE, U., PALLAGHY, C. K., WILLERT, K. VON: Microautoradiographic investigations of water lily (*Nymphaea*) leaves with special reference to the effect of poly-L-lysine. J. Membrane Biol. **4**, 395—407 (1971).

LÜTTGE, U., WEIGL, J.: Mikroautoradiographische Untersuchungen der Aufnahme und des Transportes von $^{35}SO_4^{--}$ und $^{45}Ca^{++}$ in Keimwurzeln von *Zea mays* L. und *Pisum sativum* L. Planta (Berl.) **58**, 113—126 (1962).

LÜTTGE, U., WEIGL, J.: Der Ionentransport in intakten und entrindeten Maiswurzeln. Ber. dtsch. bot. Ges. **77**, 63—70 (1964).

LÜTTGE, U., WEIGL, J.: Zur Mikroautoradiographie wasserlöslicher Substanzen. Planta (Berl.) **64**, 28—36 (1965).

OSMOND, C. B., LÜTTGE, U., WEST, K. R., PALLAGHY, C. K., SHACHER-HILL, B.: Ion absorption in *Atriplex* leaf tissue. II. Secretion of ions to epidermal bladders. Austral. J. biol. Sci. **22**, 797—814 (1969).

PALADE, G. E.: A study of fixation for electron microscopy. J. Exp. Med. **95**, 285—297 (1952).

PELC, S. R.: The stripping-film technique of autoradiography. Internat. J. Applied Radiation and Isotopes. **1**, 172—177 (1956).

PETERS, A., HINDS, P. L., VAUGHN, J. E.: Extent of stain penetration in sections prepared for electron microscopy. J. Ultrastructure Research **36**, 37—45 (1971).

RECHENMANN, R. V., WITTENDORP, E., GIELINK, A. J., RINGOET, A.: Localisation du calcium au niveau des chloroplastes de la feuille d'avoine (*Avena sativa* L.) par une méthode traceautoradiographique a très haute résolution. C. R. Acad. Sc. Paris, Série D **269**, 1525—1527 (1969).

REYNOLDS, E. S.: The use of lead citrate at high pH as an electronopaque stain in electron microscopy. J. Cell Biol. **17**, 208—213 (1963).

RINGOET, A., ZEEUW, D. DE: Discrimination between calcium and strontium in oat plants. II. Relation to the differential chemical binding of both ions. Z. Pflanzenphysiol. **59**, 249—257 (1968).

RINGOET, A., RECHENMANN, R. V., GIELINK, A. J.: Calcium accumulation in and possible excretion through guard and accessory cells of stomata and through bulliform cells. Z. Pflanzenphysiol. **64**, 60—64 (1971a).

RINGOET, A., RECHENMANN, R. V., GIELINK, A. J., WITTENDORP, E.: Calcium localization in chloroplasts of the oat leaf by high-resolution track-autoradiography. Z. Pflanzenphysiol. **64**, 80—84 (1971b).

RINGOET, A., SAUER, G., GIELINK, A. J.: Phloem transport of calcium in oat leaves. Planta (Berl.) **80**, 15—20 (1968).

ROGERS, A. W.: Techniques of Autoradiography. Amsterdam: Elsevier 1967.

SATTER, R. L., MARINOFF, P., GALSTON, A. W.: Phytochrome controlled nyctinasty in *Albizzia julibrissin*. II. Potassium flux as a basis for leaflet movement. Amer. J. Bot. **57**, 916—926 (1970)

SAWHNEY, B. L., ZELITCH, I.: Direct determination of potassium ion accumulation in guard cells in relation to stomatal opening in the light. Plant Physiol. **44**, 1350—1354 (1969).

Seelmann-Eggbert, W., Pfennig, G.: Radionuklid-Tabellen. Bundesministerium für Wiss. Forschung: Fränk. Ges. Druckerei Würzburg 1964.

Shone, M. G. T., Clarkson, D. T., Sanderson, J.: The absorption and translocation of sodium by maize seedlings. Planta (Berl.) **86**, 301—314 (1969).

Stösser, R.: Autoradiographische Lokalisierung von ^{45}Calcium im Xylem bei Fruchtstielen von *Prunus avium* L. Z. Pflanzenphysiol. **62**, 387—392 (1970).

Taugner, R., Hole, H., Grigoleit, G., Wagenmann, U.: Herstellung geeigneter Gefrierschnitte zur Autoradiographie der Niere in einer Kühlkammer mit eingebautem Mikrotom. Naunyn-Schmiedeberg's Arch. Pharmak. exp. Path. **234**, 330—335 (1958a).

Taugner, R., Wagenmann, U., Hole, H., Grigoleit, G., Droh, R., Gillissen, J.: Serienmäßige Herstellung von Gefrierschnitt-Autoradiogrammen mit optimalem Kontakt. Naunyn-Schmiedeberg's Arch. Pharmak. exp. Path. **234**, 336—342 (1958b).

Weigl, J., Lüttge, U.: Mikroautoradiographische Untersuchungen über die Aufnahme von $^{35}SO_4^{--}$ durch Wurzeln von *Zea mays* L. Die Funktion der primären Endodermis. Planta (Berl.) **59**, 15—28 (1962).

Weigl, J., Lüttge, U.: Die Ionenaufnahme durch die Luftwurzeln von *Epidendrum*. Protoplasma (Wien) **60**, 1—6 (1965).

Ziegler, H., Lüttge, U.: Die Salzdrüsen von *Limonium vulgare*. II. Die Lokalisation des Chlorids. Planta (Berl.) **74**, 1—17 (1967).

Ziegler, H., Weigl, J., Lüttge, U.: Mikroautoradiographischer Nachweis der Wanderung von $^{35}SO_4^{--}$ durch die Tertiärendodermis der *Iris*wurzeln. Protoplasma (Wien) **56**, 362—370 (1963).

Chapter 4

Microautoradiography of Water-Soluble Organic Compounds

W. Eschrich and E. Fritz
Forstbotanisches Institut der Universität
Göttingen, W.-Germany

Contents

4.1. *Introduction*

This chapter deals with microautoradiographic methods for the localization of labeled water-soluble organic compounds within plant tissues. Organic molecules occurring in plant cells are composed mainly of H, C, N, O, P, and S. From nitrogen and oxygen no radioactive isotopes with sufficiently long half-life are available for use as tracers. $^{32}PO_4^{\equiv}$ and $^{35}SO_4^{=}$, when occurring as monoesters, can be precipitated as lead and barium salts respectively, as described in Chapters 3 and 7. Procedures treated in this chapter deal mainly with ^{3}H- and ^{14}C-labeled compounds.

The techniques described in this chapter cover a wide range of applicabilities. They may be used to detect the distribution pattern of labeled compounds, assimilates as well as extraneous substances. The resolution extends from gross autoradiographs down to microautoradiographs of cell organelles.

4.2. *General Considerations*

The labeled solutes which shall be localized in tissues or cells are soluble in water. Any fixation, whether aqueous or gaseous, will break the membraneous borders of the cytoplasmic compartments causing leakage of the substances in question. Therefore the cell water must be solidified by freezing. Quick and deep freezing is essential to prevent formation of ice crystals which otherwise would damage cytoplasmic structures and disrupt delicate cell walls (see for instance Fig. 4.7). The electron microscopic freeze-etching technique (Moor, 1964) has convincingly demonstrated that structures at the submicroscopic level can be preserved very well by freezing.

For first orientations the frozen specimen may be sectioned with a cryotome, and the sections exposed to the photographic emulsion while frozen (4.3).

For more accurate resolution, the solidified water must be removed. This can be done by freeze-substitution (4.4.4) or by freeze-drying (4.4.2). Plant tissues, unlike animal tissues, are difficult to lyophilize because of the high water retention by capillaries of the cell walls. Conventional freeze-dryers generally are not able to remove completely the capillary-bound water. Nevertheless it must be removed when the specimen is to be embedded in paraffin or resin. Infiltration with an organic solvent preceeding embedding in most cases displaces (xylene) or dissolves (ether, chloroform) the remnants of water. Application of pressure (vacuum infiltration, displacement by vapor pressure) generally cannot be avoided (4.4.3).

The embedded specimen can be sectioned by conventional methods. Handling of the sections is difficult because stretching on water causes dissolution and displacement of the labeled substances. Several "dry" methods have been used (4.4.5) with more or less success. A compromise method which has several advantages is the method of hydrophobizing the specimen by siliconization during embedding (4.4.8).

For detection of radioactivity various photographic emulsions have been used. Stripping film (Kodak AR 10) as well as liquified emulsions (Eastman-Kodak NTB-2, NTB-3, Ilford G-5, K-2, K-3, K-5) give satisfactory results (see also Chapter 5.6.3).

Preceeding the development of the exposed photolayer, provision must be made to fix the sections in place. This is extremely difficult in the case of paraffin embedded material. The sections must be freed of paraffin to allow the developer to come in contact with the emulsion. In addition, removal of the paraffin is necessary for microscopic investigation of the finished microautoradiograph. In most cases some sections "swim off" during development or fixation. This never occurs with resin-embedded material, because the transparent resin must not be removed from sections for microscopic observation.

4.3. *Cryotome Techniques*

4.3.1. Gross Autoradiography

For examination of gross distribution of tracers in branches, roots and trunks of trees, cross sections are freeze-dried, glued on cardboard and exposed to X-ray film (Hull and Leonard, 1964). The techniques used for gross autoradiography on X-ray film are described by Crafts and Yamaguchi (1964) and Eschrich (1966).

The specimen is frozen in crushed dry ice and then sectioned with a cryotome. Cryotome sections, depending on the species, must be 15 to 30 or more μm thick, to reduce curling or wrinkling during the next step, i.e. freeze-drying. The sections may be flattened during drying with a weight of cooled sintermetal, preferably polished stainless steel. The dried section is then placed in a moist atmosphere for about 30 min and glued on cardboard. Any non-acid glue can be used. The cardboard with the sections is flattened overnight in a press and is then covered directly with the X-ray film, again under pressure. Medical X-ray film with blue background is preferred because of better contrast in reproductions. It usually has photo emulsion on both sides which also facilitates handling in the dark room.

Exposure at room temperature in a light-tight cabinet can be completed after 2 days when ^{14}C is used as the tracer. Overexposure leads to a dark background [1].

Control sections containing no tracer should be made to discriminate between background and radioactivity-caused blackening. The exposed X-ray film then is taken off the cardboard and developed in normal X-ray film developer.

4.3.2. Microautoradiography

Several attempts have been made to use cryotome sections for obtaining microautoradiographs with sections and photoemulsion in permanent contact. The results depend greatly on the contact between the section and the film which in turn depends on the flatness of the sections. Whereas cryotome sections of animal tissue flatten "by themselves" while drying, plant tissue sections often wrinkle because of unequal drying rates between cell walls and cytoplasm. Woody tissue seems to dry more uniformly than parenchymatous tissue.

The specimen holder of the cryotome is precooled in a freezer, and a piece of fresh tissue is mounted on the holder using an interface of wet filter paper. Tissue and holder are wrapped with thin plastic sheet and the whole device is frozen in liquid isopentane or propane (4.4.1). Before installing the holder on the cryotome, the plastic is cut away.

Sectioning is performed in the dark room illuminated by indirect red safety light. The section is picked up with a precooled slide coated with stripping film (Kodak AR 10), photo emulsion facing the section (APPLETON, 1964). For removal of the section from the knife, the slide must be slightly warmer than the knife but certainly not above the freezing point.

Sections also may be transferred from the knife to the slide by using a cooled forceps or a brush. KOLLMANN (1965) arranges the sections on a slide covered with filter paper and presses it on a slide precoated with photo emulsion (see also Chapter 3, Fig. 3.8).

Exposure is carried out in a freezer. For development, the preparations are brought to room temperature and dried by a gentle current of air. To keep the sections in position, the slides are coated with either an 8—10% gelatine solution or the slide is dipped in a solution containing cellulose acetate (4.4.6). LÜTTGE and WEIGL (1965) coated the slides with stripping film and then added an overlayer of eggwhite-glycerol (1 : 1), diluted 1 : 10

---

1 Coniferous wood also contains compounds which react with X-ray film to produce pseudoautoradiographs. HULL and LEONARD (1964) wrapped the section with Saran Wrap (Dow Chemical Comp.) to eliminate this artifact.

with water. The cryotome sections were picked up on the slide and flattened by a drop of liquid paraffin in hexane (30 : 1). After exposure, the paraffin was removed by bathing in xylene and ethanol. The dried preparations were dipped in a 0.1% gelatine solution and dried again to prevent the sections to "swim off" during processing the photo-emulsion.

Cryotome techniques are best elaborated for animal tissues. Sections as thin as 1 μm can be obtained provided the knife temperature is below —60 °C (for literature pertaining to these techniques see SCHULTZE, 1969, p. 44). For plant tissue, in contrast, this technique is utilized to a much lesser extent, probably due to the presence of cell walls. Cryosections of plant tissue can seldom be cut thinner than 10 μm.

4.4. *Embedding Techniques*

Paraffin as well as synthetic resin are utilized as embedding media. Despite many disadvantages of paraffin it must be used if large sections are necessary.

The steps of embedding procedures are outlined in the following survey:

Paraffin technique:		resin technique:	
freezing	(4.4.1)	freezing	(4.4.1)
freeze-drying	(4.4.2)	freeze-drying	(4.4.2)
or freeze-substitution with		fixation	(4.4.7)
intermedium	(4.4.4)	infiltration with intermedium	(4.4.3)
embedding in paraffin	(4.4.3)	siliconization	(4.4.8)
sectioning	(4.4.5)	embedding in resin	(4.4.9)
stretching	(4.4.5)	polymerization	(4.4.9)
dry mounting on photo emulsion	(4.4.5)	sectioning	(4.4.10)
exposure	(4.4.6)	stretching on water	(4.4.10)
deparaffinizing	(4.4.6)	staining	(4.4.10)
processing	(4.4.6)	coating with photo emulsion	(4.4.11)
staining		exposure	(4.4.11)
enclosure		processing	(4.4.11)
		enclosure	(4.4.11)

In the following subsections (4.4.1—11) the single steps of preparations will be described.

4.4.1. Freezing

Quick and deep freezing is essential to prevent formation of destructing ice crystals. Most commonly, pieces of tissue cut as small as possible are frozen in organic solvents or gases of low melting point, which are cooled with liquid nitrogen (b.p. —195.8 °C). Isopentane (m.p. —159.9 °C; b.p. 27.9 °C) and propane (m.p. —187.7 °C; b.p. —42.1 °C) are most frequently used. Some authors use isopentane containing 8% methylcyclohexane.

For cooling isopentane a suitable amount of liquid in a beaker is lowered into a Dewar flask [2]) containing liquid nitrogen. Occasional stirring with a glass rod will provide a homogeneous slurry which should have a temperature of —165 to —170 °C. The beaker then is lifted and its content stirred to get a low-viscosity slurry.

Liquid propane is produced by conducting the gas from the tank through a rubber tube into a copper tubing coiled around a beaker [3]). The end of the copper tubing is bent to open into the beaker. Both beaker and copper coil are immersed in liquid nitrogen. The propane coming from the tank will condense and drop into the beaker. The beaker must be kept in liquid nitrogen until propane starts to solidify on the walls. All operations should be carried out in a hood.

The tissue, wrapped with gauze or thin plastic wrap, is placed in the freezing-solution with a forceps. Freezing is accelerated by moving the tissue. The frozen tissues may be kept 24 hrs at —25 °C or 48 hrs at —50 °C before recrystallization begins (WINCKLER, 1970).

If the whole plant must be frozen, which is necessary for studies on translocation because sieve tube content is displaced if the bundle is cut, the plant is frozen with crushed dry ice (CRAFTS and YAMAGUCHI, 1964; ESCHRICH, 1966). It is not necessary to bury the whole plant under dry ice, but covering a towel over the plant will accelerate freezing. Most plants are completely frozen within 5 min. The parts of the plant to be embedded are sawed out with an ampoule saw, precooled on dry ice. Sawing by slow movements will prevent local thawing caused by frictional heat. Tissue frozen with dry ice should be dried or freeze-substituted without delay.

4.4.2. Freeze-Drying

The principle of freeze-drying is to remove the tissue water by sublimation. Several freeze-dryers are commercially available. Suitable equipment provides a vacuum of 10^{-3} Torr and a low condenser temperature (at least —54 °C, better —70 °C).

---

2 Dewar flasks, which have been used previously to store dry ice or mixtures of dry ice and organic solvents will break when filled with liquid nitrogen. Do not dispose liquid nitrogen down a drain! Pour it on the floor!

3 To prevent breaking of the copper tube while bent, it must be heated vigorously with a burner and suddenly chilled with water shortly before bending.

A simple device may be constructed as follows (Fig. 4.1): A high vacuum-proved desiccator (d) is put in a deep freezing compartment (f) and is connected by wide vacuum tubing to a water vapor trap (t), cooled by dry ice/acetone, and a vacuum pump (p). The frozen tissues are placed in the precooled desiccator together with a small reserve of dry ice. The pump must run continuously during the operation which will take 2—14 days depending on the species and the size of the specimen. More sophisticated devices have aluminum tanks instead of glass desiccators and are equipped with a vacuum gauge.

Plant tissue which has been lyophilized in this way can be sufficiently embedded only after previous infiltration with an organic intermedium (4.4.3).

Direct infiltration with paraffin requires a special device which lyophilizes the tissue under a current of gas (JENSEN and KAVALJIAN, 1956; BRANTON and JACOBSEN, 1961; Jensen's Botanical Histochemistry, 1962, p. 119). A comparable device has been used by ESCHRICH (1966) (Fig. 4.2). BOLLI (1967) has described a simpler apparatus. This kind of lyophilizing under gas flow (moving gas method) provides an equal desiccation of the tissue. A low current of CO_2 or N_2, containing 2.6—4.0 μg of water per liter (for determination of the water content see JENSEN, 1962, p. 123) prevents a high vacuum, but dries the tissue block equally. If a block of frozen tissue is exposed to high vacuum it will dry superficially, and the inter-

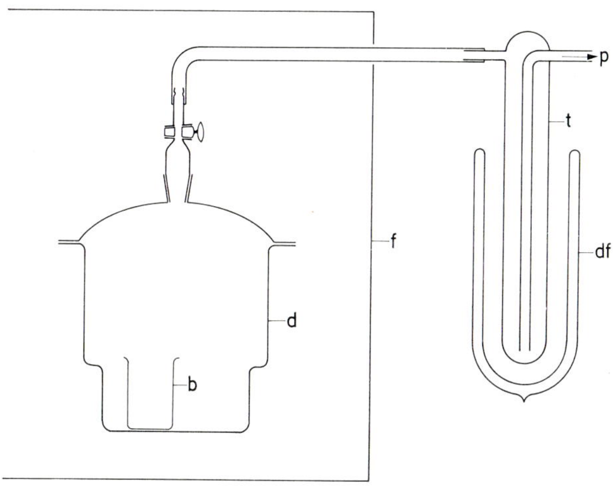

Fig. 4.1. Simple device for freeze-drying plant tissue blocks. The desiccator (d) containing the frozen specimen in a beaker (b) is situated in a deep-freeze compartment (f) and evacuated over a cooling trap (t) by a vacuum pump (p). The Dewar flask (df) is filled with a mixture of dry ice and acetone

fibrillar spaces of the cell walls will collapse, forming a nearly impenetrable covering around the block. Thus, tissue blocks seemingly dry and brittle often contain a core of ice inside.

The time necessary for drying varies from 2—10 days depending on the kind of tissue, the vacuum, and the intensity of the gas flow. Evacuation should continue when the gas flow is turned off. A final vacuum of 10^{-3} Torr kept for several hours is required for complete infiltration with paraffin. Satisfactory results obtained with the gas flow method are published by MORTIMER (1965), ESCHRICH and STEINER (1967), BOLLI (1967) and HEYSER et al. (1969).

4.4.3. Infiltration

Infiltration of freeze-dried plant tissue with molten paraffin seems to be possible only if the gas flow technique is utilized for drying. After the final evacuation the Dewar flask (Fig 4.2) which keeps the specimen frozen, is replaced by a water bath, which is warmed until the paraffin melts. To prevent foaming, the paraffin must have been previously degased. Then the specimen holder is lowered into the paraffin. Dry pieces which are satisfactorily infiltrated will sink to the bottom. Pieces remaining on the upper side of the holder sometimes will sink down when the vacuum is broken suddenly. These pieces however contain usually gas bubbles and are unsuitable for serial sections. For final embedding, the specimen holder is transferred to a beaker containing molten paraffin and is opened. The pieces are removed with a hot forceps and are put in suitable capsules for solidifying.

Tissues, which have been freeze-dried without gas flow in most cases must be infiltrated with an intermedium prior to embedding.

Xylene (mixture of isomers), chloroform, and ether (diethyl ether) have been used as intermedia.

STERLING and CHICHESTER (1956) infiltrated freeze-dried fruit tissue (peach) with a 10% solution of parlodion (celloidine) in ethanol/ether (1:1). The parlodion was hardened in chloroform. After transfer into xylene, the tissue was embedded in paraffin. This procedure should prevent wrinkling of the sections, but the alcohol-ether intermedium certainly will dissolve some of the water-soluble material.

For embedding in resin the procedure described previously (FRITZ and ESCHRICH, 1970), with some modifications, is recommended:

Small blocks (4 by 4 mm) of freeze-dried tissue are fixed (4.4.7), kept in an evacuated desiccator for 2 days over $CaCl_2$, and transferred into the bowl (b) of the device for vacuum infiltration (Fig. 4.3a). If several different blocks are infiltrated together, they must be labeled and wrapped separately

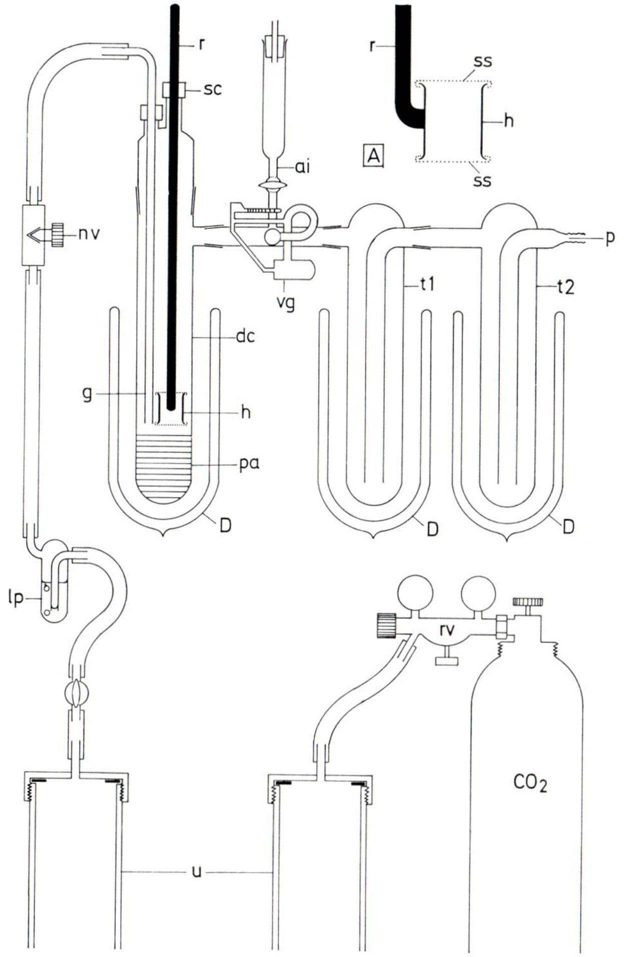

Fig. 4.2. Device for freeze-drying plant tissue blocks under gas flow. The frozen tissue in its holder (h), which is closed with fabric of stainless steel (ss in Fig. 4.2A), can be lifted up and down by the glass rod (r) through a screw cap with teflon packing (sc) (Quickfit, Schraubkappenverschluß). The drying chamber (dc) contains previously degased paraffin (pa). During drying a gentle stream of gas, escaping from the glass tube (g) rounds the holder. The gas comes from a tank (CO_2) with reducing valve (rv), runs through an U-shaped wide tube filled with $CaCl_2$(u) (2.6—4.0 μg water/liter), and bubbles through liquid paraffin (lp). It can be regulated by a needle valve (nv). The drying chamber (dc) is connected by a side arm with two cooling traps (t1, t2) and a vacuum pump (p). Between drying chamber and the first cooling trap (t1) a vacuum gauge (vg) (according to Brunner. A. Pfeiffer GmbH, Wetzlar BRD) is attached (NG 14.5/35). An inlet for dry air (ai) allows the sudden break of the vacuum. Dewar flasks (D) are filled with mixtures of dry ice and acetone. Trap t2 must be the coolest

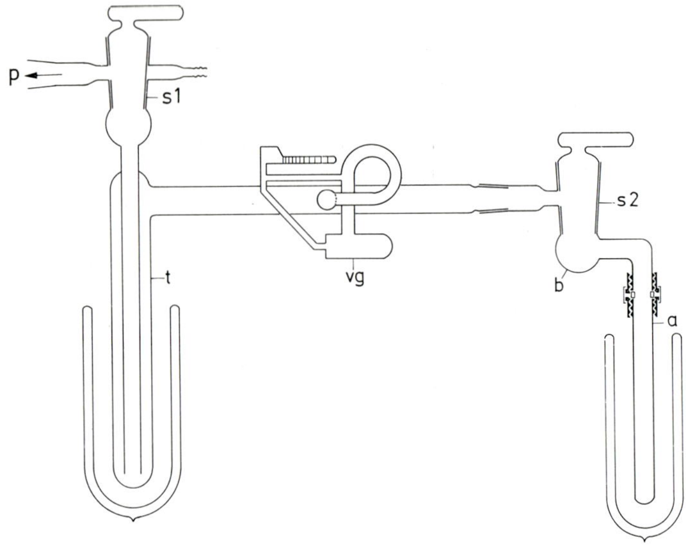

Fig. 4.3a. High vacuum apparatus for infiltration of freeze-dried plant tissue with intermedia. The ampoule (a) (Sovirel, 22 mm i.d. 200 mm long), containing the intermedium, is connected to a one-way stopcock (s2) (Brand, 12 mm side-boring). The tissue is placed in the bowl (b) of the stopcock, which is connected with a cooling trap (t), bypassing the vacuum gauge (vg). A two-way stopcock (s1) connects the cooling trap with a mercury diffusion pump (p) (Schott, 3Q 6) and a water aspirator

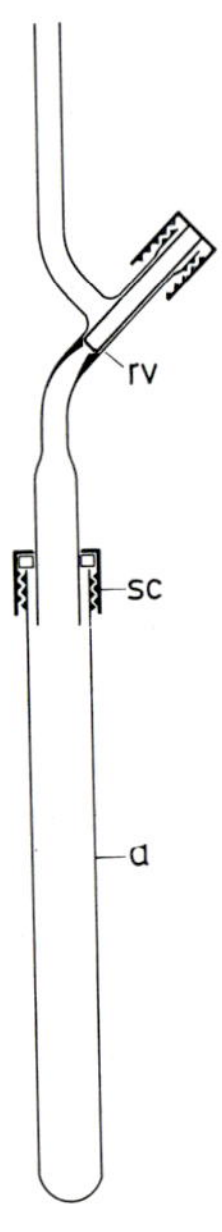

Fig. 4.3b. Ampoule (a) (see Fig. 4.3a) equipped with Rotaflo valve (rv) (Quickfit) and gas-tight srew cap (sc) (Sovirel) for evacuation with a water aspirator preceeding the final infiltration under vapor pressure

with fabric of stainless steel (V2 A-S Drahtgewebe, Gebr. Kufferath, 5161 Mariaweiler, BRD). The ampoule (a), connected with the one-way stopcock (s2) by a double-screw fitting (Quickfit, Sovirel) contains sodium-dried ether. The connected device is evacuated with a water aspirator until 1—2 ml of ether are vaporized. Then the ampoule and the cool trap (t) are cooled with liquid nitrogen. After crystallization of the ether, a mercury vapor diffusion pump (Schott, Mainz, 3 Q6) is turned on to evacuate to 10^{-2} Torr. Then the Dewar flasks are removed for degasing the ether. As soon as foaming and bubbling stops, the ether first is crystallized again and then the cool trap is cooled with liquid nitrogen. After approximately 2 hrs of evacuation, the stopcock (s2) is closed, pumps are turned off, and the Dewar flask under the ampoule is removed. The stopcock (s2) is disconnected from the vacuum gauge arm, the tissue blocks are dropped on top of the melting ether and the system remains closed for 2 days, at room temperature.

Some tissue blocks may have sunken by this time.

Infiltration sometimes will take 6 days and in extreme cases up to 10 days. Pieces which are extremely slowly infiltrated, are usually extremely dry.

To promote infiltration, the ampoule (a) is disconnected from the stopcock (s2) at room temperature, provided with an evacuating device (Sovirel SVL 22 and Quickfit Rotaflo valve. rv in Fig. 4.3b), and evacuated carefully with a water aspirator to remove the air which has invaded during disconnection. After this the valve is closed, and the ampoule placed in a water bath of 50 °C for about 2 days. The high vapor pressure of the ether (b.p. 37 °C!) in almost all cases completes infiltration. The ampoule must be cooled to room temperature before it is opened under the hood.

Recent experiments have shown that evacuation with a diffusion pump and crystallization of the ether can be omitted. This simplifies the procedure and is as follows: Freeze-dried and fixed (4.4.7) tissues are encaged and placed in the bowl (b) of the infiltrating device Fig. 4.3a. The ether in the side arm ampoule (a) then is evaporated at room temperature with a water aspirator until the level has dropped about 2 cm. During the first few minutes evacuation must be regulated by hand to prevent strong foaming. After this the stopcock (s2) is closed, side arm and stopcock are disconnected and the encaged specimens are dropped into the ether. After 2—3 hrs the side arm ampoule (a) is disconnected from the stopcock (s2) at room temperature, provided with a Rotaflo valve (rv in Fig. 4.3b), and carefully evacuated with a water aspirator until some ether has evaporated. Then, the valve is closed and the ampoule kept for 6 days in a water bath of 50 °C. Cooling the ampule periodically with ice water promotes infiltration.

4.4.4. Freeze Substitution

Instead of freeze-drying and infiltrating a frozen tissue, it can be dehydrated by exchange of the molecules of water by those of an organic solvent. This process may take a considerable period of time and is usually carried out at approximately —30 °C.

Blank et al. (1951) substituted frozen animal tissue, labeled with ^{32}P, by propylene glycol at —20 °C and embedded in a mixture of water-soluble polyethylene glycols (Carbowax).

Neeracher (1965) substituted ^{3}H-labeled tissue of *Zea mays* with absolute ethanol at —70 °C. He found that no difference in the distribution of the tracer occurred as compared with freeze-dried and infiltrated tissue. The anatomical structure of the tissue was kept better by freeze-substitution.

Lüttge and Weigl (1965) used ether to substitute the water at —30 °C. The essential steps of preparation are the following:

Ether, freed of peroxides and water by treatment with Al_2O_3 is precooled to —30 °C. $^1/_{10}$the volume of Al_2O_3 and the frozen specimen are added. The mixture is shaken at —30 °C for 2—6 days. By this treatment the tissue-water is absorbed by the desiccant. The dehydrated tissue blocks are transferred into methylbenzoate and embedded in paraffin. The authors found that this method prevents shrinking and rupture of the tissue.

Note added in proof:
Fisher (1972) and Fisher and Housley (1972) recently have compared freeze-substitution and freeze-drying for preparation of microautoradiographs of water-soluble compounds. The results show some advantages of the freeze-substitution method.

References

Fisher, D. B.: Artifacts in the embedment of water-soluble compounds for light microscopy. Plant Physiol. **49**, 161—165, 1972

Fisher, D. B., Housley, T. L.: The retention of water-soluble compounds during freeze-substitution and microautoradiography. Plant Physiol. **49**, 166—171, 1972.

4.4.5. Sectioning and Mounting Paraffin-Embedded Tissues

Specimens embedded in paraffin, whether directly or by previous infiltration with an intermedium, are sectioned with a rotary microtome. The usual procedure of stretching single sections or parts of a ribbon on water can not be applied if microautoradiographs of water-soluble compounds are desired. These compounds will leach out immediately when the section is brought into contact with water.

Stretching on mercury or aniline has been tried without success.

Holt et al. (1949) introduced the technique in which the block of paraffin is covered with a drop of molten wax after each cut. The next section then remains flat and is easy to transfer because of the relatively thick layer of wax. Gallimore et al. (1954) fastened the ribbon of paraffin sections on the small borders of a rectangular hole in a brass plate. By careful heating over a hot plate the ribbon stretches and eventually hangs down through the hole. Nevertheless, these sections are not flat and will wrinkle when pressed on the photo emulsion. A similar device using a teflon-coated wire has been described by Eschrich and Steiner (1967). Other authors omit stretching and arrange the sections on a slide or cover glass precoated with the photo emulsion and press them firmly with a teflon-coated roller or sheet. Lüttge and Weigl (1965) stretch the sections on the photo layer by dropping ether or xylene on them. This will raise the background of the photo emulsion (Neeracher, 1965). Branton and Jacobson (1962) recommend heating the slide with the photo emulsion to 45 °C when pressing the sections on it.

It is obvious that none of these "dry-mounting" techniques is satisfactory. In most cases, the sections will wrinkle and have unequal contact with the photolayer.

In every case, sections must be applied on top of the photo emulsion, because the paraffin must be dissolved before processing the emulsion. The techniques of coating slides with photo emulsion (stripping film as well as liquid emulsions) are essentially the same as described in other chapters of this book. If stripping film is used, the side of the film facing the supporting glass plate must also face the slide. If in the finished microautoradiographs the emulsion is wanted on top of the section, a cover glass is fixed to the slide by a film of glycerol and the assembly is coated with the photo emulsion. For enclosure, the cover glass is lifted with a razor blade from the slide and mounted section down on a new slide (see Chapter 3, Fig. 3.8).

4.4.6. Exposure and Development

The slides with the sections on the photo emulsion normally are exposed prior to deparaffinizing. Neeracher (1965) suggests to remove the paraffin before exposure to raise the yield of recorded radioactivity.

Exposure is carried out in any light-tight box in which a bag of dust-free desiccant is enclosed. The box must be sealed with an adhesive tape to exclude moisture. Most authors prefer to expose in a refrigerator.

The time of exposure depends on several factors, mainly on the amount and specific activity of the labeled substances. For ^{14}C-labeled material seldomly less than 2 weeks are necessary to develop a latent image. Sources with low specific activity sometimes need several months of exposure time.

Exposure for more than 6 months will raise the background and occlude the microautoradiographs. Being generous with the number of slides prepared is suggested, because a good many of them will be used for checking purposes.

Preceeding development the paraffin must be removed. If this is performed by bathing in xylene, many sections will swim off when the slide is developed, fixed and watered. To prevent this, the slide must be protected by a thin film of material which allows penetration of xylene and dissolved paraffin, as well as developer and fixative, and which provides sufficient support to keep the sections in place. Most recommended is the method of Branton and Jacobson (1962), who dipped the slide with the paraffin sections on it into a solution of 0.5% cellulose acetate in ethylmethylketone to which 10% of acetone had been added. After a short drying period the sections are deparaffinized in 2 changes of xylene. Remnants of xylene must be removed completely before development. Best is to dry the slides in a desiccator over solid paraffin for 48 hrs. Silver grains lining the cell walls, are usually an artifact produced by remnants of xylene.

For development, Kodak D 19b-developer is most recommended. The recipe is included in each box of AR 10-stripping film plates. The slides are developed for 5 min at 20 °C. For fixation, an acid fixative (1—2 min fixation time) is preferred. After watering (10 min) the slides should be rinsed in distilled water. Slow drying in a desiccator over dust-free silicagel is recommended, because drying in a current of warm air may lead to the formation of fissures.

Sometimes silver grains are very small. In this case, development should be prolonged up to 20 min (see Fig. 4.6). Staining and enclosure of paraffin-sectioned microautoradiographs is as usually in microtechnique. Some dyes stain the gelatine of the photo emulsion. Congo red has been proved satisfactory (Eschrich, 1966).

4.4.7. Fixation of Freeze-Dried Tissue

Formaldehyde has been used by Benditt et al. (1965) to fix ^{14}C-labeled proline in freeze-dried fibroblasts. This method has been adapted by Fritz and Eschrich (1970) to fix water-soluble amino acids in freeze-dried plant tissue. The dry tissue blocks are trimmed to convenient sizes by utilizing splinters of razor blades. This procedure should be performed in a room with less than 50% relative humidity to keep the tissue dry and brittle.

After trimming, the tissues are exposed for 8 hrs to vapors of paraformaldehyde, which is spread on a petri dish standing on top of a layer of dry $CaCl_2$ in a desiccator. The closed desiccator is put in an oven of 60 °C. After fixation, the desiccator is opened under the hood, and the fixed tissues are stored in an evacuated desiccator over $CaCl_2$ before infiltration.

Vapor escaping from dry paraformaldehyde penetrates the tissue and fixes amino acids on solid parts of the cells or on polymers of the aldehyde. BENDITT et al. (1965) found that fixation with formaldehyde keeps 71—79% of the label in the tissue when the latter was treated with water.

Two pieces of the same petiole of *Vicia faba* containing ^{14}C-labeled phenylalanine, one fixed with vapor of paraformaldehyde and the other unfixed, were compared as final microautoradiographs. We observed that the radioactivity of the unfixed tissue sections was washed out when the (resin-embedded) sections came in contact with water. The fixed sections showed the expected distribution of the tracer (Fig. 4.4).

Fixation with vapors other than formaldehyde apparently has not yet been investigated. Labeled sugars are not fixed by vapor of paraformaldehyde. To prevent leakage of sugars out of resin-embedded tissues, siliconization (4.4.8) is recommended.

4.4.8. Siliconization

Sugars and other compounds which are not fixed by vapor of paraformaldehyde are washed out when the sections come into contact with water. STIRLING and KINTER (1967) added silicon oil (1%) to the embedding resin, and found that the leakage of ^{14}C-labeled galactose from sections of animal tissue was prevented. This method also has been proved valid for plant tissue (FRITZ and ESCHRICH, 1970).

The tissues, infiltrated with ether and transferred to dry xylene, are placed for 24 hrs in xylene containing 0.2% silicon oil. Any kind of silicon oil, even high-vacuum grease, can be utilized. A part of the silicon oil will separate as small droplets when the tissue is transferred into the xylene-resin mixture 2 : 1 (4.4.9).

The effect of siliconization can be deducted from the following experiment: Two pieces of a petiole (*Vicia faba*), which was labeled with ^{14}C-phenylalanine, were processed for microautoradiographs. The tissue in Fig. 4.4a and b has been fixed with vapor of paraformaldehyde (4.4.7) but was not siliconized, the tissue in Fig. 4.5a and b has been only siliconized. The figures show no difference in the amount and distribution of the label. Siliconization may be sufficient to prevent leakage of all kind of water-soluble compounds, whereas formaldehyde-fixation seems to fix only substances containing amino- or imino groups (BENDITT et al., 1965). Practically, the additional fixation with formaldehyde is only a safety factor, but it is recommended since it does not involve any technical difficulties.

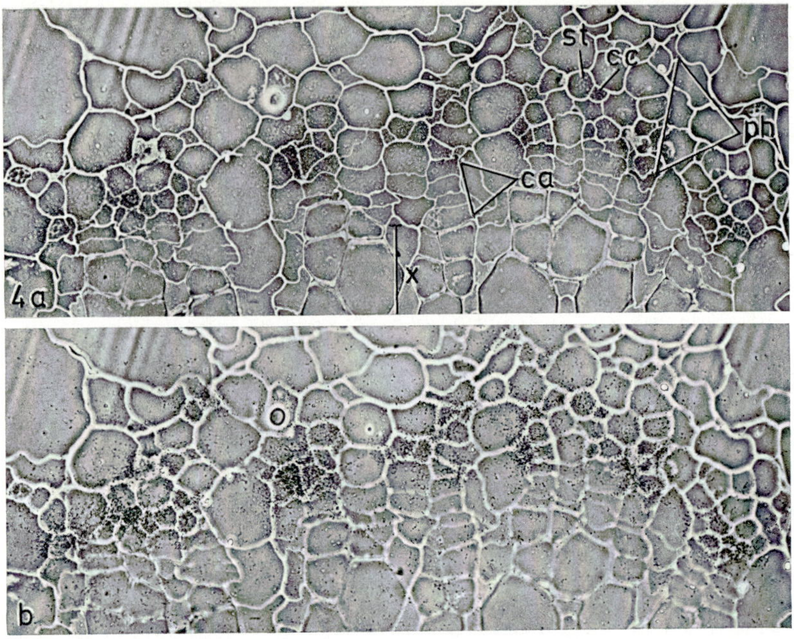
st
cc
ph
ca
x
4a
o
b

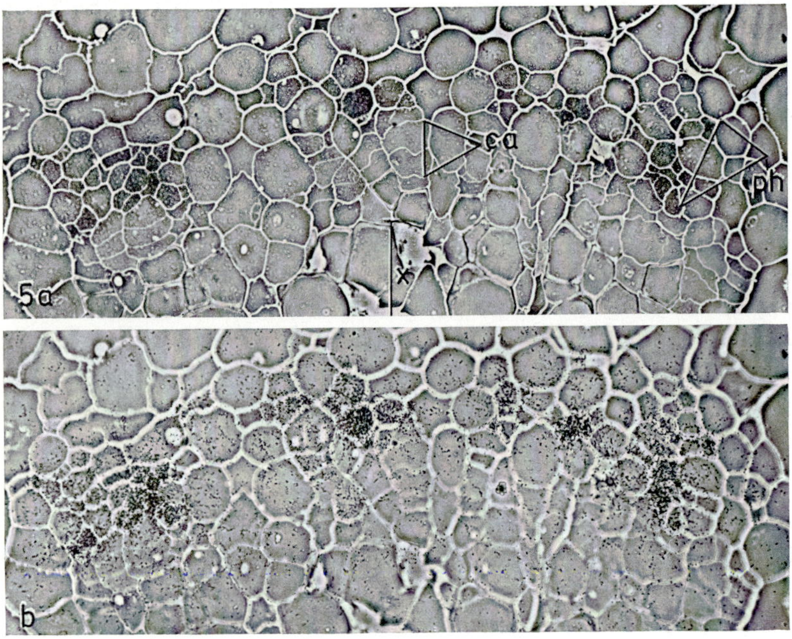
ca
ph
x
5a
b

4.4.9. Embedding in Resin and Polymerization

Specimens infiltrated with ether are transferred into the following mixtures, remaining 3 hrs in each: ether/xylene 2 : 1, ether/xylene 1 : 1, ether/xylene 1 : 3, and finally sodium-dried xylene. Then siliconization (4.4.8) is carried out. For embedding, Durcupan ACM (Fluka) has been used by FRITZ and ESCHRICH (1970).

The siliconized tissues are transferred stepwise into mixtures of xylene/resin mixture, which are kept at 50 °C. The resin mixture is composed of 10 ml epoxyresin, 10 ml hardener and 0.15 ml softener (all ingredients are provided by Fluka). The specimens remain 5—8 hrs in xylene/resin mixture 2 : 1 (v/v), 5—8 hrs in xylene/resin mixture 1 : 1, and 48 hrs in xylene/resin mixture 1 : 3.

During step 1 and 2, the vials remain closed. For step 3 small open capsules of folded aluminum foil (heavy duty), about 15 by 20 mm in size are convenient. After 24 hrs, the odor of xylene can no longer be detected. After 48 hrs, all xylene should have escaped.

Next, 2 changes of pure resin mixture are passed, each for 48 hrs at 50 °C. Occasional stirring with a needle is recommended. For the final step, 0.3 ml activator are added to 20 ml of resin mixture. The tissues remain in this mixture 15 days; 3 days at 40 °C, 3 days at 50 °C, 3 days at 60 °C and 6 days at 70 °C. During the first 3—5 hrs, occasional stirring is of importance.

The slow raising of temperature is necessary to get sufficient contact between the polymerizing resin and the cell walls.

During polymerization, sometimes small gas bubbles form in the inner part of the tissue blocks. The origin of these bubbles is not yet well understood. It cannot be water vapor, because bubbles occur after infiltration with xylene as well as with ether, and dry ether should remove completely all water. It also cannot be xylene, because Durcupan ACM polymerizes free of bubbles in the presence of relatively great amounts of xylene. Therefore it is supposed that volatile components of the resin mixture remain captured inside the block. This may happen when the activator invades incompletely. Few and small bubbles, in general, do not cause much trouble but greater amounts of bubbles indicate incomplete polymerization resulting in partly soft resin blocks and bulging sections.

Fig. 4.4 and 4.5. Microautoradiographs of two cross sections of a vascular bundle from the petiole of a *Vicia faba* plant. The leaf blade has been fed for 3 hr with 9 μc D,L-phenylalanine-1-^{14}C. Unstained sections. Exposure time 42 days. 6 min developed at 20 °C. Phase-contrast photographs. 310 ×. Upper pictures (a) focused on section, lower pictures (b) focused on the photo emulsion. ca, cambial region; cc, companion cell; ph, phloem; st, sieve tube; x, xylem. 4.4, tissue fixed with paraformaldehyde, but not siliconized. 4.5, tissue siliconized, but not fixed

As suggested by LUFT (1961), we tried to polymerize the xylene/resin mixture 1 : 3 by addition of 1.5% activator. These blocks hardened completely and free of bubbles, because the relatively high xylene content provided low viscosity and facilitated penetration of the activator. Unfortunately, sections obtained from these blocks showed rupture of the resin from the cell walls.

A possibility to reduce bubble formation may be to use a resin mixture which polymerizes in UV-light without a catalyst. The use of low viscosity resins, such as the mixture of epoxy-resins recommended by SPURR (1969), also could solve the "bubble problem". For these materials, no experience for microautoradiography of water-soluble substances and siliconized tissue is available as yet.

4.4.10. Sectioning and Staining Resin-Embedded Material

Resin-embedded tissue, which is fixed with vapor of paraformaldehyde and siliconized can be sectioned, stretched and stained by conventional methods.

The trimmed blocks are sectioned with glass knives on an ultramicrotome. The thickness of the sections depends on the nuclide used for labeling. For recording ^{3}H-label on Kodak AR 10 stripping film (which has a 5 μm photolayer), thin sections of about 0.5 μm are suitable. The yield of silver grains produced by ^{14}C-electrons is greatest if relatively thick sections are used, but thick sections cause enlargement of the halo of scatter-radiation. Most recommended are sections of 1—1.5 μm thickness.

The sections floating on dust-free bidistilled water in the trough are transferred with an eyelash to a drop of water on a clean slide and, if serial sections are wanted, arranged in a circle around the drop. Fifteen to 20 sections can be arranged in a drop of 10—15 mm diameter. Between the first and the last section of the series a dot is scraped on the slide with a diamond-tipped marker. The slide must be cleaned with nitric acid, washed, dried and provided with a film of gelatine before the sections are transferred to it. Gelatine sheets are superior to powdered gelatine, because they provide a harder surface. The "subbing solution" in which the dry slide must be dipped is composed of 0.5% gelatine (Merck No. 4072) and 0.05% chromealum in bidistilled water. The gelatine sheets are soaked in water at 60 to 70 °C overnight and filtered hot before the chromealum is added.

The sections are stretched in an oven at 50 °C for 1—2 min. Vapor of chloroform supplied by a small cotton ball held over the sections will complete stretching. After this, the sections are allowed to dry completely at 50 °C.

Staining the cell walls is necessary for microscopy under bright-field illumination. For phase-contrast microscopy (see Figs. 4.4, 4.5) the staining procedure can be omitted. A series of recipes for staining methacrylate-embedded sections has been given by FEDER and O'BRIEN (1968). Some of these dyes stain the Durcupan. According to our experience, a 1% aqueous solution of gentian violet B (C.I. No. 42535) works satisfactorily. The slides are kept for 15 min at room temperature in the staining solution, washed for 5 min in running tap water, dipped in bidistilled water and dried in a dust-free cabinet. Cell walls are usually somewhat overstained, but they will be destained to a violet color during development and fixation.

4.4.11. Coating with Photoemulsion, Exposure, Development and Enclosure

For obtaining uniform microautoradiographs which allow a semiquantitative estimation of the density of the silver grains, a defined thickness of the photo emulsion is necessary. This is best provided by stripping films. Kodak AR 10 stripping film is recommended. It can be stored refrigerated for up to 6 months. Coating the slides with stripping film is performed in the dark room under indirect safety light (Philips red light PF 712E). The pieces of stripping film are floated, photo emulsion facing the water, on a dust-free surface of distilled water until soaked. The film is picked up with the slide in the usual way (see Chapter 3., Fig. 8), dried for 24 hrs in the dark over silicagel and exposed in light-tight boxes to which a small bag of silicagel is enclosed. The boxes are closed air-tight with adhesive tape and kept in a refrigerator at 4 °C.

No schedule for estimating the exposure time is available. The best procedure is to expose 3 series of slides, and then develop them at different intervals.

Development in Kodak D 19b developer at 20 °C, fixation, watering and drying at a dust-free place are essentially as described elsewhere.

For enclosure under cover glass, it is noted that unstained preparations to be observed by phase-contrast illumination give good contrast using gum Arabic dissolved in 33% potassium acetate as mounting medium (ESCHRICH and CURRIER, 1964). Gelatine mixtures, which must be warmed for mounting, will weaken the supporting gelatine of the photolayer and may disarrange the silver grains. For stained microautoradiographs any benzene-soluble resin is applicable.

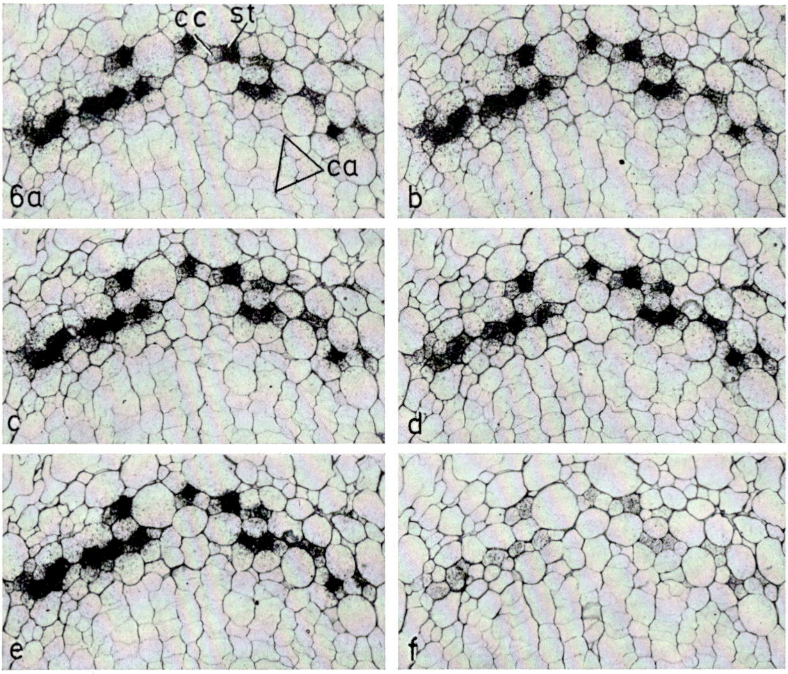

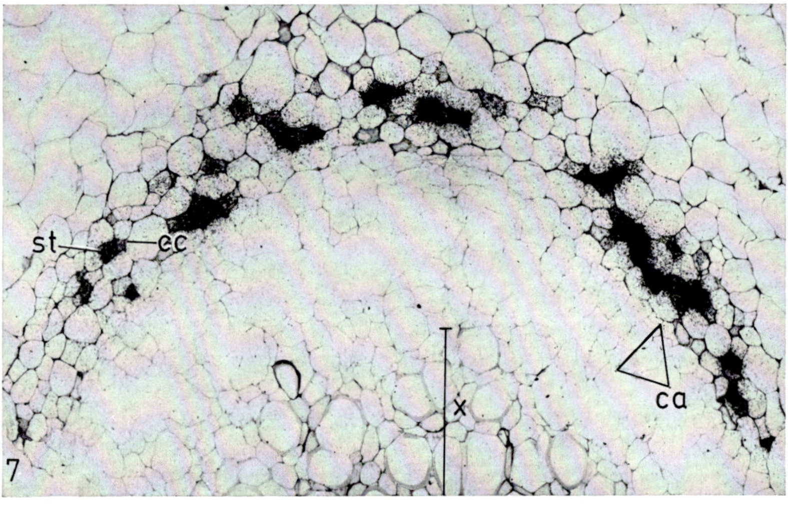

Fig. 4.6. Microautoradiographs of a series of 1 μm thick cross sections of a phloem bundle of the stem of *Vicia faba*. The blade of the superimposed leaf has been exposed for 35 min to 130 μc $^{14}CO_2$. Sections stained with gentian violet. Exposure time 131 days. Bright-field illumination. 190 ×. a—e, subsequent sections. 20 min developed at 20 °C. f, 6 min developed at 20 °C

4.5. *Examples*

A survey of the literature pertaining to the microautoradiography of water-soluble organic compounds in plant tissues shows, that most publications contain much more technical comments than results which can be considered as useful. The techniques have been described in the previous sections (4.3 and 4.4), examples of good microautoradiographs are to find in PERKINS et al. (1959), WEBB and GORHAM (1964), MORTIMER (1965), SCHMITZ and WILLENBRINK (1967, 1968), TRIP and GORHAM (1967), ESCHRICH and STEINER (1967), HEYSER et al. (1969), GEIGER et al. (1969), SCHMITZ (1970) and FRITZ and ESCHRICH (1970). In all except the last cited publications, techniques for embedding in paraffin have been used. "Dry methods" seem to be essential for obtaining useful microautoradiographs, but "dry method" means: tricky details, crumpled sections, lots of waste and occasional success. This raises the question whether the relatively-time-consuming paraffin techniques better are replaced by the more-time-consuming resin-embedding technique, which guarantees nearly 100% successful preparations (FRITZ and ESCHRICH, 1970).

Some examples may show, what can be expected from resinembedded material. Microautoradiographs in Fig. 4.4 to 4.8 where obtained from vascular bundles of *Vicia faba* plants, which had been fed either with 9 μc D,L-3-phenylalanine-1-^{14}C (Fig. 4.4, 4.5), or 130 μc $^{14}CO_2$ (Figs. 4.6, 4.7, 4.8), over a single leaf. Resin sections were 1 μm thick and were exposed on Kodak AR 10 stripping film.

Fig. 4.7 shows a cross section of a vascular bundle with radioactivity only in sieve elements and companion cells of the metaphloem. This tissue was frozen in dry ice and shows ruptures in the cambial region caused by ice crystals. Fig. 4.6 shows a series of subsequent cross sections. It demonstrates that serial sections of great conformity are to obtain with resin-embedded material. The first 5 sections were developed for 20 min, the last only for 6 min. Serial longitudinal sections through a single sieve element are shown in Fig. 4.8. All photographs described so far have been taken from stained preparations with bright-field illumination. Fig. 4.4 and 4.5 show unstained microautoradiographs photographed with phase-contrast illumination. Sufficiently enlarged phase-contrast pictures are suitable for grain counting.

Fig. 4.7. Part of a cross section of a vascular bundle from a *Vicia faba* stem, which was labeled as in Fig. 4.6. Freezing in dry ice caused rupture of cell walls in the cambial region (ca) by ice crystals. Exposure time 216 days. Stained with gentian violet. 20 min developed at 20 °C. 220 ×. ca, cambial region; cc, companion cell; st, sieve tube; x, metaxylem

Fig. 4.8. Microautoradiographs of a series of three (a—c) subsequent longitudinal 1 μm sections of the phloem. Same vascular bundle as in Fig. 4.7. Stained with gentian violet. Exposure time 32 days. 6 min developed at 20 °C. Bright-field illumination. 255 ×. cc, companion cell; se, sieve element; sp, sieve plate

References

Appleton, T. C.: Autoradiography of soluble labeled compounds. J. Roy. Micr. Soc. **83**, 277—281 (1964).

Benditt, E. P., Martin, G. M., Platter, H.: Application of freeze-drying and formaldehyde-vapor fixation to radioautographic localization of soluble amino acids. Symposia of the Internat. Soc. for Cell Biology, vol. **4**, p. 65—75, ed. C. P. Leblond, K. B. Warren. New York: Academic Press 1965.

Blank, H., McCarthy, P. L., Lamater, E. D. de: A non-vacuum freezing-dehydrating technic for histology, autoradiography and microbial cytology. Stain Technol. **26**, 193—197 (1951).

Bolli, H. K.: Absorption und Transport markierter Aminosäuren durch Wurzeln von *Zea mays*. Ber. Schweiz. Bot. Ges. **77**, 61—102 (1967).

Branton, D., Jacobson, L.: Freeze-drying of plant material. Exp. Cell Res. **22**, 559—568 (1961).

Crafts, A. S., Yamaguchi, S.: The autoradiography of plant material. Calif. agric. exp. sta. extension service, manual 35, 143 p. Berkeley, Calif. 1964.

Eschrich, W.: Translokation ^{14}C-markierter Assimilate im Licht und im Dunkeln bei *Vicia faba*. Planta **70**, 99—124 (1966).

Eschrich, W., Currier, H. B.: Identification of callose by its diachrome and fluorochrome reactions. Stain Technol. **39**, 303—307 (1964).

Eschrich, W., Steiner, M.: Autoradiographische Untersuchungen zum Stofftransport bei *Polytrichum commune*. Planta **74**, 330—349 (1967).

Feder, N., O'Brien, T. P.: Plant microtechnique: some principles and new methods. Amer. J. Bot. **55**, 123—142 (1968).

Fritz, E., Eschrich, W.: ^{14}C-Mikroautoradiographie wasserlöslicher Substanzen im Phloem. Planta **92**, 267—281 (1970).

Gallimore, J. C., Bauer, E. C., Boyd, G. A.: A non-leaching technic for autoradiography. Stain Technol. **29**, 95—98 (1954).

Geiger, D. R., Saunders, M. A., Cataldo, D. A.: Translocation and accumulation of translocate in the sugar beet petiole. Plant Physiol. **44**, 1657—1665 (1969).

Heyser, W., Eschrich, W., Evert, R. F.: Translocation in perennial monocotyledons. Science **164**, 572—574 (1969).

Holt, M. W., Cowing, R. F., Warren, S.: Preparation of radioautographs of tissues without loss of water-soluble P^{32}. Science **110**, 328—329 (1949).

Hull, R. J., Leonard, O. A.: Physiological aspects of parasitism in mistletoes (*Arceuthobium* and *Phoradendron*). I. The carbohydrate nutrition of mistletoe. Plant Physiol. **39**, 996—1007 (1964).

Jensen, W. A.: Botanical histochemistry. San Francisco and London: W. H. Freemann Co. 1962.

Jensen, W. A., Kavaljian, L. G.: The cytochemical localization of ascorbic acid in root tip cells. J. Biophys. Biochem. Cytol. **2**, 87—92 (1956).

Kollmann, R.: Zur Lokalisierung der funktionstüchtigen Siebzellen im sekundären Phloem von *Metasequoia glyptostroboides*. Planta **65**, 173—179 (1965).

Lüttge, U., Weigl, J.: Zur Mikroautoradiographie wasserlöslicher Substanzen. Planta **64**, 28—36 (1965).

Luft, J. H.: Improvements in epoxy resin embedding methods. J. Biophys. Biochem. Cytol. **9**, 409—414 (1961).

Moor, H.: Die Gefrierfixation lebender Zellen und ihre Anwendung in der Elektronenmikroskopie. Z. Zellforsch. **62**, 546—580 (1964).

Mortimer, D. C.: Translocation of the products of photosynthesis in sugar beet petioles. Canad. J. Bot. **43**, 269—280 (1965).

NEERACHER, H.: Transportuntersuchungen an *Zea mays* mit Hilfe von THO und Mikroautoradiographie. Ber. Schweiz. Bot. Ges. **75**, 303—346 (1966).

PERKINS, H. J., NELSON, C. D., GORHAM, P. R.: A tissue-autoradiographic study of the translocation of C^{14}-labeled sugars in the stems of young soybean plants. Canad. J. Bot. **37**, 871—877 (1959).

SCHMITZ, K.: Untersuchungen zur funktionellen Anatomie des Leitgewebesystems im Blatt von *Pelargonium zonale*. Planta **92**, 208—221 (1970).

SCHMITZ, K., WILLENBRINK, J.: Histoautoradiographischer Nachweis ^{14}C-markierter Assimilate im Phloem. Z. Pflanzenphysiol. **58**, 97—107 (1967).

SCHMITZ, K., WILLENBRINK, J.: Zum Nachweis tritiierter Assimilate in den Siebröhren von *Cucurbita*. Planta **83**, 111—114 (1968).

SCHULTZE, B.: Autoradiography at the cellular level. Physical techniques in biological research, vol. **3**, part B, ed. A. W. POLLISTER. New York and London: Academic Press 1969.

SPURR, A. R.: A low-viscosity epoxy resin embedding medium for electron microscopy. J. Ultrastruct. Res. **26**, 31—43 (1969).

STERLING, C., CHICHESTER, C. O.: Autoradiography of water-soluble materials in plant tissues. Stain Technol. **31**, 227—230 (1956).

STIRLING, C. E., KINTER, W. B.: High-resolution radioautography of galactose-^{3}H accumulation in rings of hamster intestine. J. Cell Biol. **35**, 585—604 (1967).

TRIP, P., GORHAM, P. R.: Autoradiographic study of the pathway of translocation. Canad. J. Bot. **45**, 1567—1573 (1967).

WEBB, J. A., GORHAM, P. R.: Translocation of photosynthetically assimilated C^{14} in straight-necked squash. Plant Physiol. **39**, 663—672 (1964).

WINCKLER, J.: Zum Einfrieren von Gewebe in Stickstoff-gekühltem Propan. Histochemie **23**, 44—50 (1970).

Chapter 5

Microautoradiography of Organic Compounds Insoluble in a Wide Range of Polar and Non-polar Solvents

R. G. Herrmann and W. O. Abel
Max-Planck-Institut für Pflanzengenetik
6802 Ladenburg/Rosenhof, W.-Germany

Contents

Abbreviations used

DNA	deoxyribonucleic acid
RNA	ribonucleic acid
CDP	cytidine-5'-diphosphate
dCDP	deoxycytidine-5'-diphosphate
TdR	thymidine
TTP	thymidine-5'-triphosphate
FUdR	5'-fluorodeoxyuridine
DMSO	dimethyl sulfoxide
PCA	perchloric acid
TCA	trichloroacetic acid

5.1. *Introduction*

Physiologically active and structurally central substances, such as DNA, RNA, proteins, polysaccharides and lipids are usually after fixation insoluble in a wide range of solvents. The presence, the localization and the quantity of radioactive isotopes incorporated into these compounds, can be demonstrated by autoradiography in organelles, cells or tissues, and often simultaneously in one preparation. Accordingly, autoradiography may be used to solve problems involving the site, initiation, duration and rate of synthesis of substances. Often it is even possible to investigate the type of synthesis, e.g. DNA replication or repair synthesis. The redistribution of substances from parent to daughter cells can also be analysed. Autoradiography is especially useful for demonstrating and measuring small amounts of radioactivity, and is therefore an independent though complementary method to liquid scintillation and gas flow counting techniques. The sensitivity is sufficient to detect a few decays per week.

Using the technique of autoradiography one must consider the properties of the isotope, the nature of the nuclear emulsion and the treatment of the specimen, to obtain the desired information. The following review of the problems, which have been investigated by autoradiography, is primarily limited to botanical objects. We have, however, not reviewed the total literature on this technique (for earlier literature see JENSEN, 1962).

5.2. *Application of Autoradiography*

DNA: Many papers have dealt with the course of nuclear DNA synthesis during division cycles of vegetative and generative cells (mitosis: HOWARD and PELC, 1951; QUASTLER and SHERMAN, 1959; WIMBER, 1960, 1961, 1966; FRIEDBERG and DAVIDSON, 1970 — meiosis: TAYLOR, 1953, 1959; WIMBER and PRENSKY, 1963; RILEY and BENNET, 1971). The mode of duplication and the chromosome structure have also been investigated (TAYLOR, WOODS and HUGHES, 1957; TAYLOR, 1958a; PEACOCK, 1963; WOLFF and LUIPPOLD, 1964; MARTIN, 1968; NELSON, 1969; COMINGS, 1971; BURKHOLDER and MUKHERJEE, 1971), and sister-strand exchange was demonstrated (TAYLOR, 1958b; PEACOCK, 1963). In the last few years the exchange patterns during meiosis were also analysed (TAYLOR, 1965; MOENS, 1966; DOUGLAS and KROES, 1969; JONES and CRAIG-CAMERON, 1969; CHURCH and WIMBER, 1969).

The low amounts of DNA in chloroplasts and mitochondria could be demonstrated autoradiographically (chloroplasts: STOCKING and GIFFORD, 1959; WOLLGIEHN and MOTHES, 1964; HERRMANN, 1970; mitochondria: GREEN and BARR, 1970). DNA synthesis in nuclei and chloroplasts was found to occur at various times in the cell cycle of *Euglena* (COOK, 1966).

DNA can be localized by both incorporation of a labeled precursor during the S-phase or by detection of stable complexes formed between a labeled compound and DNA or nucleoprotein. These methods have been useful in demonstrating small amounts of DNA or nucleoproteins possessing unique properties. With 3-H-actinomycin D it was possible to label DNA in chloroplasts (WOODCOCK and BOGORAD, 1970) and nucleoli (EBSTEIN, 1969); differences in actinomycin binding between eu- and heterochromatin were also demonstrated (BERLOWITZ, PALOTTA and SILBEY, 1969). Furthermore it is possible to hybridize DNA with isolated labeled DNA or RNA *in situ* (DNA: PARDUE and GALL, 1970; ARRIGHI, HSU, P. SAUNDERS and G. F. SAUNDERS, 1970; RNA: BUONGIORNO-NARDELLI and AMALDI, 1970; JOHN, BIRNSTIEL and JONES, 1969). In this way it was possible to delimit the origin of the satellite DNA in mice in the heterochromatin adjacent to the centromere (PARDUE and GALL, 1970).

RNA: The occurrence of RNA synthesis and its intracellular localization were established after relatively few investigations in root tip cells (JENSEN, 1957; SISKEN, 1959; WOODS and TAYLOR, 1959; WOODARD, RASCH and SWIFT, 1961), in meiocytes (TAYLOR, 1958c, 1959; DE, 1961; ALBERTINI, 1965; DAS, 1965; SAUTER, 1969), in tapetal cells (TAYLOR, 1959; ALBERTINI, 1965; SAUTER, 1969), in development of pollen, and also during pollen tube growth (STEFFENSEN, 1966; MASCARENHAS, 1966).

Relations between RNA synthesis and morphological chromosome changes in the polytene chromosomes found in the suspensor cells of *Phaseolus vulgaris* were readily established by NAGL (1969, 1970), after inhibition of RNA synthesis by actinomycin-D and rifamycin.

Proteins: The existence of overall differences in the protein synthetic activity was initially investigated during mitosis by JENSEN (1957); CLOWES (1958); WOODARD, RASCH and SWIFT (1961); PATE and O'BRIEN (1968) and during meiosis by TAYLOR (1958c, 1959); ALBERTINI (1967); SAUTER (1969). Later, specific proteins were investigated, including for example the synthesis of basic histones or acidic proteins in the cell nucleus (WOODARD, RASCH and SWIFT, 1961; CAVE, 1966, 1968; PRESCOTT, 1966; BLOCH, MCQUIGG, BRACH and WU, 1967; SCHNEIDER and RIEKE, 1967; SHAPIRO and POLIKARPOVA, 1969). These might possibly participate in gene regulation. The labeling of chromosomal protein was also adapted to monitor protein transfer by the sperm nuclei into the eggs (STEFFENSEN, 1965).

In addition, the protein synthesis in the nucleoli (BIRNSTIEL and HYDE, 1963; MATTINGLY, 1963; CHOUINARD and LEBLAND, 1967), the syntheses of membranes in chloroplasts (GOLDBERG and OHAD, 1970), and the problem of whether plant cell walls contain any protein (SAVADA and CHRISPEELS, 1969), were investigated autoradiographically.

A recent method permits localization of proteins with specific metabolic function (BARNARD, 1970). This "labeled inhibition method" utilizes labeled compounds which form an insoluble complex with specific enzymes or enzyme groups; e.g. the localization of esterases with the substrate 3-H-di-isopropylfluorophosphate (OSTROWSKI et al., 1964; BARNARD and OSTROWSKI, 1964). Antigens may be localized by similar techniques (OSTROWSKI, et al., 1970).

Polysaccharides: In the area of polysaccharide metabolism advances concerning the site of pectin synthesis and cellulose synthesis have been the focus of some investigations (BÖHNER, 1958; SETTERFIELD and BAILEY, 1959; WARDROP, 1959; WARDROP and HARADA, 1965; JOHNSON, 1965; JOHNSON and GIBSON, 1966a, b; ROBERTS and BUTT, 1967, 1968, 1969). In other studies lignification was analysed by REZNIK (1960) and FREUDENBERG et al. (1955). Because of its higher resolution, electron microscopic autoradiography is the preferable method for such studies (see Chapt. 6).

Lipids: To our knowledge, plant lipids have not been studied by autoradiography; thus we can only refer to work on animals (COPE and WILLIAMS, 1968; DERMER, 1968; STEIN et al., 1969).

Among many other investigations which used autoradiography, those which analysed the binding of phytohormones to chromatine should be noted (LIAO and HAMILTON, 1966).

5.3. *Selection of the Isotope*

The radioactive isotopes used for the autoradiography of "insoluble" compounds almost exclusively emit β-particles (electrons). Among the β-emitters 3-H and 14-C are most commonly used, while 35-S, 32-P, 33-P and 125-I are used less often.

For a given isotope β-particles take up various amounts of energy during radioactive decay, up to a characteristic maximal amount (E_{max}). The initial energy of a β-particle, its distance and track length in the nuclear emulsion and also the number of silver grains per track are correlated (LEVI et al., 1963).

Isotopes with electrons of E_{max} up to about 0.040 MeV (3-H, 125-I) are considered as weak, those with an E_{max} more than 0.400 MeV (32-P) as strong emitters (cf. Table 1 in Chapt. 3). β-particles with low initial energies (about 0.02 MeV) penetrate about 3 μm into the emulsion and produce tracks up to about 4 grains (most electrons from 3-H reach only 1—2 µm into the emulsion and produce only 1—2 grains). Electrons with up to about 0.200 MeV (14-C, 35-S, 33-P) travel distances up to 25 µm and maxima of about 80 grains per track were reported (most electrons emitted from 14-C reach 5—10µm into the emulsion). For electrons with a still higher initial energy (32-P) the distance of the rays from the source of decay can amount to several mm.

Regarding these characteristics one may distinguish two basic types of autoradiographs, a) grain density autoradiographs and b) track autoradiographs. Each type is applicable to different questions (cf. 5.6.3.2, see Fig. 5.1).

5.4. *Experimental Prerequisites for Incorporation*

5.4.1. Selection of Precursors

All isotopes mentioned above can often be incorporated into known positions in organic compounds. In turn, these may be utilized to generate biological structures. The labeled molecules were used either in the metabolism of the compounds being investigated (5.2), in which case they are designed as precursors, or they react in the living or fixed specimen with compounds to give an insoluble complex (5.2, see also Chapt. 3.4.2).

Precursors must be selected carefully. Wherever possible they should represent the intermediate closest to the examined end product. They should be readily incorporated by the cell, and they should not be appreciably metabolized into other substances.

If precursors labeled with different isotopes are available, then the desired information determines the choice of the precursor (5.6.3.2). By using 3-H labeled compounds localizations are possible even within cell organelles (Fig. 5.2). 3-H labeled precursors should be used fresh, since self-radiolysis is possible (WAND et al., 1967). Furthermore, the isotope can be exchanged with adjacent substances (5.12.3). Both possibilities diminish the specificity of incorporation.

DNA: The use of 32-P (orthophosphate) has been discontinued, except in special cases, because of the inspecificity of incorporation, the low resolution which is obtainable, and the low half-life of the isotope. Usually one uses 3-H thymidine, typically labeled in the 6-position of the thymine ring. The use of methyl labeled thymidine can be disadvantageous since the side chain may be metabolized. However, ring-labeled 3-H-TdR may contain a quantity of thymidine labeled at the 5-methyl group (see data sheets of the manufacturers). Thymine is less suitable as precursor for DNA than thymidine (FRIEDKIN, TILSON and ROBERT, 1954; REICHARD and ESTBORN, 1951).

However, it is sometimes difficult or impossible to label DNA with thymidine. When the phosphorylation enzymes (e.g. thymidine kinase; FINK and FINK, 1962) needed for its activation in the cell are lacking, it cannot be used as a precursor, as for instance in fungi (CHAKRABATORY and LORING, 1960) or presumably in various algae (STEFFENSEN and SHERIDAN, 1965; SAGAN, 1965; SAGAN et al., 1965). Labeled purine nucleosides are then used (ROSENBERGER and KESSEL, 1968). A high internal pool of thymidine or high catabolism of thymidine might represent other difficulties. In the first case at least, blockage of the cells own thymidine synthesis by the addition of 5-fluorodeoxyuridine may be advantageous (10^{-6} to 5×10^{-3} M, COHEN et al., 1958).

125-I-deoxyuridine can also be used instead of 3-H- and 14-C-labeled thymidine (ROGERS, 1967).

For *in vitro* incubation of isolated cell organelles labeled deoxyribonucleotide triphosphates must be selected (SCOTT et al., 1968).

RNA: In spite of the ubiquitous presence of uridine (uracil) in RNA, various findings indicate that more specific labeling of RNA can be obtained with 3-H- and 14-C-cytidine. Uridine may also be used for the biosynthesis of thymidine and is involved in more biosynthetic pathways than is cytidine (TSAI and AXELROD, 1965; CLEAVER, 1967; ROSS and COLE, 1968). Regardless of the precursor base used, it is essential to test the specificity of incorporation in RNA by treatment with DNase and RNase (5.12.3).

Proteins: In labeling proteins the specific protein type must be considered in selecting the precursor amino acids. Arginine or lysine (labeled with 3-H or 14-C) are often selected for basic proteins (histones), tryptophan, valine or 35-S-labeled methionine for acidic proteins (Lit. see 5.2).

Polysaccharides: 3-H-D-glucose has often been used, but has been replaced recently by 14-C-6D-glucuronate (ROBERTS and BUTT, 1968) or 14-C-1D-galactose (ROBERTS and BUTT, 1969) in order to attain a higher incorporation specificity.

Lipids: For lipids, various fatty acids, glycerin and steroids were used as precursors (Lit. see 5.2).

5.4.2. Amount of Precursor for Incubation/Incubation Time

The amount of precursor to be given depends on several factors. Basically, one must attain sufficient incorporation levels to insure adequate autoradiography, but these levels should not generate serious radiation damage in the living specimen (ZAJICEK and GROSS, 1964). The exposure time depending on the type of emulsion should not exceed $^{1}/_{2}$ to 1 year, or three half-life times in short lived isotopes (cf. Chapt. 3.3.1). The precursor concentration in the incubation medium should be correspondingly higher if the time of incubation is short or if the concentration of the substance to be labeled in a cell is low, or if the internal cold pool of precursor is large. The incorporation efficiency in the presence of a high internal precursor pool can be increased by inhibiting the synthesis of precursor in the cell (5.4.1.).

Table 5.1. Relationship between the Amount of Radioactive Precursor, Incubation Time

Sub-stance	Organelle (kind of tissue)	Precursor	Spec. act. (C/mM)
DNA	*Nuclei*		
	root	3-H-thymidine	3.6
	root	3-H-thymidine	3.6
	meiospore	3-H-thymidine	6.4
	gonial cells	3-H-actinomycin D	3.4
	Chloroplasts		
	leaf	3-H-thymidine	3.0
	leaf	3-H-thymidine	5.0
	cell	3-H-guanine	0.33
	isolated, fixed	3-H-actinomycin D	—
	isolated	3-H-TTP	4.8
RNA	*Nuclei*		
	root	14-C-orotic acid	0.7 (mC)
	meiospore	3-H-cytidine	1.9
	meiospore	3-H-uridine	7.9
	meiospore	3-H-cytidine	1.0
	suspensor	3-H-uridine	6.1
Protein	root	3-H-lysine	0.25
	root	3-H-lysine	0.2
	root	3-H-arginine	0.1
	root	3-H-arginine	1.05
	root	3-H-arginine	1.83
	root	3-H-tryptophane	0.5
	root	14-C-leucine	0.37 (mC)
	meiospore	3-H-leucine	0.5
Poly saccharide	*Cell wall*		
	root	14-C-galactose	3.05 (mC)
	yeast	3-H-glucose	0.47

Ordinarily, low concentrations of precursors with high specific activity should be used. If high precursor concentrations are used the likelihood of precursor metabolism (5.12.3, Chapt. 6.6) and the production of other artifacts exists (5.4.3). Some data for the selection of precursors, their concentrations, incubation times in various biological specimen are contained in Table 5.1.

and Exposure Time for Different Plant Systems

Dose (μC/ml)	Time of incubation	Exposure	Reference
1	35 min	14 d	Friedberg and Davidson (1970)
0.5	30 min	3 m	Wimber (1966)
20	12 hr	1—2 m	Church and Wimber (1969)
2—50	1—4 hr	1—4 hr	Berlowitz et al. (1969)
100	8—10 hr	6 w	Wollgiehn and Mothes (1964)
10	1—72 hr	5 w	Herrmann (1970)
8	48 hr	7—8 w	Sagan et al. (1965)
25	1 hr	3—4 w	Woodcock and Bogorad (1970)
50	15 min	4—6 w	Scott et al. (1968)
5	6—12 hr	variab.	Sisken (1959)
10	1—3 d	4 w	Sauter (1969)
25	15 min	3 w	Mascarenhas (1966)
200	2 hr	2—8 w	Das (1965)
10	20; 60 min	2 w	Nagl (1969)
10	15 min	8 d	Mattingly (1963)
1	30 min	2 w—2 y	Bloch et al. (1967)
5	15 min	18 d	Mattingly (1963)
1	30 min	2 w—2 y	Bloch et al. (1967)
100	5 min	2 d	Chouinard and Leblond (1967)
10	15 min	6 w	Mattingly (1963)
2	6—12 hr	variab.	Sisken (1959)
10	2—3 hr	30 d	Sauter (1969)
0.3	3 hr	2 d	Roberts and Butt (1969)
20	14 min	10 d	Johnson (1965)

5.4.3. Methods for Incubation

To ensure normal development during the experiment, the labeled substances should be available in the culture medium to the plants, tissues, cells or organelles without disturbing normal cellular endogenous rhythms. These problems are minimal when we deal with submerged growing organisms or tissue cultures. Apical roots may be incubated in the medium according to WOLFF (1964). Several methods were developed for the incubation of pollen mother cells. MARIMUTHU (1970) threads a wool yarn through the bud. The yarn is then dipped into a solution containing the tracer. The best method seems to be the injection of labeled substance into the empty cup of the bud (SAVAGE and WIGGLESWORTH, 1971). In this way less damage should occur, as with equally efficient methods of ITO and STERN (1967) who cultured pollen mother cells in axenic media. The solution may be sterilized by passage through a bacterial filter by means of a syringe. The culture media of HOAGLAND, WHITE or BENECKE were used. With differentiated tissues, a gentle infiltration under vacuum in a large dessiccator is often utilized (HERRMANN, 1970).

Pulse labeling is performed by fixation after a short incubation (cf. Chapt. 6.6). To follow the incorporated precursor requires a chase experiment. This implies further cultivation in new media without precursor or briefly in fresh medium with excess of cold precursor (10—100-fold concentration). The cold chase technique replaces the bulk of the adsorbed radioactive precursor and simultaneously dilutes unincorporated precursor in the cell. Large doses of cold precursor, however, can lead to serious disturbances (e.g. the cytological disruption of meiosis through deoxyadenosine; ITO and STERN; cf. also 6.6). High concentration of thymidine leads to the accumulation of TTP, which blocks reduction of CDP to dCDP. This interference effect is used advantageously for synchronizing cell populations (VILLADSEN and ZEUTHEN, 1970).

5.5. *Fixation*

During fixation, in addition to good preservation of structure, one must ensure above all that (1) incorporated label is not extracted (e.g. lipids cf. DERMER, 1968); (2) non-specific linkages between unincorporated precursor and insoluble cell components do not occur, and (3) an enzymatic digestion of a fixed labeled specimen is possible as a control (5.12.3). The selection of a fixation mixture largely determines the preparation and further treatment of a specimen. Specimens are usually fixed for 24 hrs at room temperature.

To preserve and subsequently stain *nuclear structures* one can fix with Carnoy's fluid I (ethanol 96% and glacial acetic acid 3 : 1, v/v) or with ethanol, chloroform, glacial acetic acid (6 : 3 : 1, v/v/v, Carnoy's fluid II).

However, for *nuclear proteins* other fixatives must be used (CAVE, 1968) since Carnoy's fluid I and II fixatives remove arginine-containing proteins to a certain extent (ALLFREY et al., 1967;; GOROVSKI and WOODARD, 1967). Obviously, this is not the case for LILLIE's fixative (1965) which consists of 2 parts 40% formaldehyde or 25% glutaraldehyde, 1 part of glacial acetic acid and 17 parts of 10% ethanol. STEFFENSEN (1965) demonstrated that fixation with this mixture allows for controls relative to nucleases and proteases.

Other fixation mixtures are 10% formaldehyde (10 ml 40% formaldehyde and 30 ml phosphate buffer 0.025 to 0.1 M, ca. pH 7.0); 3% glutaraldehyde also in phosphate buffer (cf. also 6.7.1), and acrolein. 10% acrolein yields good fixation qualities at 0°C (FEDER and O'BRIEN, 1968). Formaldehyde and glutaraldehyde fixatives may lead to unspecific linkage of amino acids due to crosslinking properties (PETERS and ASHEY, 1967; VANHA-PERTULA and GRIMLEY, 1970; cf. also 4.4.7 and 6.7.1). Should it be impossible to avoid such fixatives, an experiment utilizing chase conditions may be useful.

For retaining the structure of isolated *chloroplasts*, 7% glutaraldehyde or 5% OsO_4 in phosphate buffer has proven satisfactory (HERRMANN, 1970; 5.8, Fig. 5.2). However, after this, control experiments, with nucleases or proteases become difficult or impossible. In contrast fixation in Carnoy's fluid I or formaldehyde allows digestion though structural detail is poorly preserved.

5.6. *Preparation of the Specimens*

5.6.1. Preparation of Slides

The slides require special cleaning and coating so that the specimen and the nuclear emulsion adhere to the glass during photographic processing and staining procedures. The coating lessens the production of artifacts during drying of the emulsion (5.12.1). Two adhesives, gelatine and albumen, are generally used for coating. The coating of the slides with albumen may be an advantage in subsequent enzymatic treatment. "Zein" is recommended as an adhesive for controls with proteases (0.5% Zein in 70% ethanol w/v; STEFFENSEN, 1965).

5.6.1.1. Cleaning of Slides

a) Place slides into glass racks and soak for about 12 hrs in acid bichromate solution (dissolve 100 g K-bichromate into approx. 800 ml demineralized water, slowly add 100 ml of conc. sulfuric acid and make up to 1.000 ml with H_2O). Rinse thoroughly in tap water, then in demineralized water, and dry under dustfree conditions.

b) We prefer extensive washing with detergents, rinsing in demineralized water, drying and treatment in boiling isopropanol/chloroform (1 : 3, v/v, 5 min, water bath about 80 °C).

The clean slides must be wetted readily by water, i.e. droplets should not form.

5.6.1.2. Recipes for Adhesives

a) Albumen/glycerol: fresh egg albumen is mixed with an equal vol. of glycerol and 2 gm Na-salicylate per 100 ml solution is added. The mixture is stirred at 40° until homogeneous (ca. 30 min). The pale solution is filtered through a glass sinter while warm. Albumen/glycerol is stable in a refrigerator for several years.

b) Gelatine/chrome alum: see 4.4.10. The solution should be made freshly every day.

5.6.1.3. Coating Slides with Adhesive

The clean slides are picked up with a clamp-forceps, dipped into gelatine-chrome alum or "Zein" solution and then dried at room temperature.

Slides for coating with albumen/glycerol are treated as follows: after dipping (best at 40°), the slide is held vertically, the adhesive is allowed to drain off, and the excess on both sides of the slide is scraped off with a ground glass slide. The slides dry at about 60° for 3 days. The adhesive layer on the slide should be as thin as possible. Since even albumen/glycerol coated slides may be kept for several years, if stored dry at room temperature, one may clean and coat a greater number and store them in plastic slide boxes. Washed and coated slides can only be touched at their edges, since the film and specimen may not adhere at the encounter point.

5.6.2. Specimen Preparation

After fixation, further treatment is determined by (1) the kind of substance (DNA, RNA etc.), (2) the fixative used, (3) the kind of isotope, (4) the method of evaluation (qualitative or quantitative) and (5) the timing of staining (before or after the coating of the preparation with emulsion).

After fixation the specimen is usually washed twice with 5% ice cold TCA for 5—10 min and then with ice-cold H_2O or buffer to remove unincorporated precursor before other control treatment (5.12.3).

Single cell or cell organelles (HERRMANN, 1970) can be dried directly on the coated slides in a cold air stream after washing out the fixative (e.g. by repeated low speed centrifugation in buffer). Thicker specimen are squashed or cut. Depending on the isotope the preparations should generally not exeed a thickness of 1—5 µm (5.6.3.2). Only with strong β-emitters like 32-P the thickness of the specimen is less critical.

The maceration, necessary for *squash preparations* is usually obtained by acidic or enzymatic hydrolysis. In DNA labeling experiments, Feulgen staining is often used before coating the sample with photographic emulsion is performed (5.11.1). The hydrolysis involved in this process softens the tissue sufficiently, so that squash preparations do not exeed specimen thickness required for autoradiography. However, hydrolysis may reduce the incorporated label by as much as 50% or more (LANG and MAURER, 1965) or it may lead to unspecific labeling (BRYANT, 1969). When labeling with purines (5.4.1) it is impossible to work with hot and diluted acid (TAMM et al., 1952).

A gentle maceration of fixed plant tissue may be obtained by treatment with pectinase (SETTERFIELD, SCHREIBER and WOODARD, 1954). Small tissue pieces are treated from 2—24 hrs with 0.5—5% pectinase which is filtered immediately after preparation and the pH adjusted to 4.0 with HCl. For a shorter maceration time, POWELL (1968) suggests the use of heat-stable pectinol 59-L (Fa. Rohm&Haas) for about 30 min at 45 °C. After this particular enzymatic method it is difficult to squash the cells following fixation with the LILLIE's-mixture (5.5) or aldehydes, even in 45% acetic acid.

The *sectioning technique*, in spite of its inherent problems is preferred in many instances, especially in the studies of RNA- and protein-labeling and also after aldehyde fixation. Freeze sectioning possesses some advantages among other sectioning techniques. A minimal section thickness can be obtained (below 1 µm; cf. 4.4.10) and the specimen may be embedded without dehydration in organic reagents, as is common in other embedding methods, for instance in glycol methacrylate or paraffin (4.4).

For freeze sectioning KNOX (1970) gives the following procedure:

a) Preparing a gelatine-anti-freeze-medium. This medium serves to surround and stabilize the tissue while sectioning but does not infiltrate the tissue. For cryostat temperatures of —10 to —15 °C, a medium of 15% gelatine (w/v) and 0.8% DMSO (v/v) or 2% glycerol or 1.5% ethyleneglycol may be used. Petri dishes with this medium are kept at 2 °C.

b) Tissue treatment: Tissue fixed in alcohols must be rehydrated in H_2O or buffer for 12—24 hrs. Meristems may be treated further immediately, while cells with vacuols must be equilibrated in 10% DMSO or 10% glycerol (DMSO 1—2 hrs at room temperature, glycerol 18—20 hrs at 37 °C). In highly vacuolated cells, an infiltration of gelatine for several hours at 37 °C may prove advantageous.

c) Preparation of tissue block: A metal rod, heated in water, is used to melt a hole into a gelatine plate (a). When the medium in the cavity has cooled down, the tissue is placed into the sol and oriented as desired. After the medium has gelled a block is cut and frozen in Freon-spray or by immersion in liquid nitrogen.

Among the resins used for embedding, the glycol methacrylate according to FEDER and O'BRIEN (1968) seems to possess several advantages (see also RUDDEL, 1967) but it cannot be removed from the sections. Though this does not interfere with subsequent staining it could possibly hinder the enzyme control. In addition, the dehydration reagents may cause changes in the specimen.

The slides ready for coating with emulsion may be marked at the corner with a diamond scribe.

5.6.3. Choice of the Nuclear Emulsion

5.6.3.1. Commonly Used Nuclear Emulsions

Two groups of techniques with many methodical variations were developed for light microscopic autoradiography, the stripping film (DONIACH and PELC, 1950; PELC, 1956) and the liquid emulsion technique (BÉLANGER and LEBLOND, 1946; KOPRIWA and LEBLOND, 1962).

Stripping films: Almost all of the experience with the stripping film technique involves stripping film AR-10, a product of Kodak. Data and handling of AR-10 are found in Chapter 3.4 and some in Chapter 5.6.3.2.

Liquid emulsions: The most widely used products are the emulsions of Ilford (Ilford Ltd. Essex, Great Britain) and Eastman Kodak (Rochester, N.Y., USA). A number of other emulsions are listed by BASERGA (1967).

Ilford emulsions are manufactured in 3 sizes of silver halide crystals G, K, L and 6 sensitivities (0—5). Seventeen of the possible combinations are obtainable.

The crystal size of the G-emulsions varies between about 0.19 and 0.36 μm (mean 0.27 μm), that for the K-emulsions lies between about 0.13

and 0.25 μm (mean 0.20 μm) and for L-emulsions between 0.07 and 0.17 μm (mean 0.14 μm). The emulsions with a sensitivity grade 0 have the lowest sensitivity. Emulsions with sensitivity grade 2 (and sometimes 1, depending on the problem) record β-particles of 3-H and 125-I, those of grade 4 (and 3) also of 14-C and 35-S, and those of grade 5 record all β-rays including those with very low ionisation energies (e.g. 32-P).

Eastman Kodak emulsions for light microscopic autoradiography with β-emitters are designated by the codes NTB, NTB 2 and NTB 3 with medium sizes of silver halide crystals of 0.29, 0.26 and 0.34 μm, resp. The NTB emulsion compares in sensitivity grade to the Ilford scale 1—2, NTB 2 to grade 3—4 and NTB 3 to grade 5. In its sensitivity AR-10 stripping film lies between grade 2 and 3 (Kodak data sheet P-64; ROGERS, 1969).

Ilford and Eastman-Kodak emulsions differ in several properties (ROGERS, 1969), e.g. viscosity (5.6.4.2), safelight requirements (5.6.4.1), sensitivity to repeated melting (5.6.4.2), sensitivity to rapid drying (5.12.1) and required conditions for exposure (5.9).

5.6.3.2. Coordination of Isotope and Nuclear Emulsion for a Given Problem

An emulsion is selected especially according to sensitivity, silver grain size and thickness of the layer. Emulsions with the lowest possible sensitivity should be used since a less sensitive emulsion has a lower tendency to create artifacts (5.12.1).

If possible, one should use the stripping film technique with AR-10. AR-10 is a well characterized product as far as sensitivity (5.6.3.1), silver grain size and emulsion thickness (which may vary somewhat above the specimen; 5.12.1) are concerned. It may be used with all above mentioned isotopes, it is suggested, however, that emulsions of higher sensitivity be used with β-emitters such as 32-P. The liquid emulsion technique which permits the possibility of varying layer thickness, sensitivity and grain size, is adaptable for special problems. If many autoradiographs of the same type are to be produced the method is especially useful because of its rapidity.

For the *localization* of radioactive compounds in a specimen one selects *grain density autoradiographs* and preferably weak β-emitters to obtain optimal resolution (5.3., Fig. 5.1). Low speed electrons lose much energy while passing through the emulsion, therefore emulsions with low sensitivity and with small silver halide crystals (about 0.1—0.2 μm) should be used, e.g. AR-10, Ilford K2, L2, L3 or Eastman-Kodak NTB. Because of the short path of these β-particles, emulsion thickness is of secondary importance. However, thickness of the specimen should not exceed 1 μm (5.8, 5.12.3).

AR-10, K3 and K4, L3, L4 as well as NTB 2 can also be used for localization problems in grain density autoradiographs of the isotopes 14-C and 35-S. When 32-P is utilized emulsions with larger crystals and still higher sensitivity are preferred (G5, K5, NTB 3). The resolution in grain density autoradiographs for the former isotopes may be improved by choosing a more sensitive emulsion combined with a reduced layer thickness of 1—3 μm (Fig. 5.1). This will enhance the detection of their low energy electrons and of the initial low energy loss of their high energy electrons. Highest resolution is obtained with thin specimen (below 1 μm), with small silver crystals and thin emulsion layers (1—2 μm). This can only be attained with liquid emulsions.

For strong β-emitters (32-P) *track autoradiographs* can give better resolution as compared to grain density autoradiographs, since the initial straight track can be extrapolated to its origin (5.8; see Fig. 3.15). The thickness of the layers, up to 150 μm are necessary for track autoradiographs (5.6.4.2), can only be produced with liquid emulsions. The silver halide

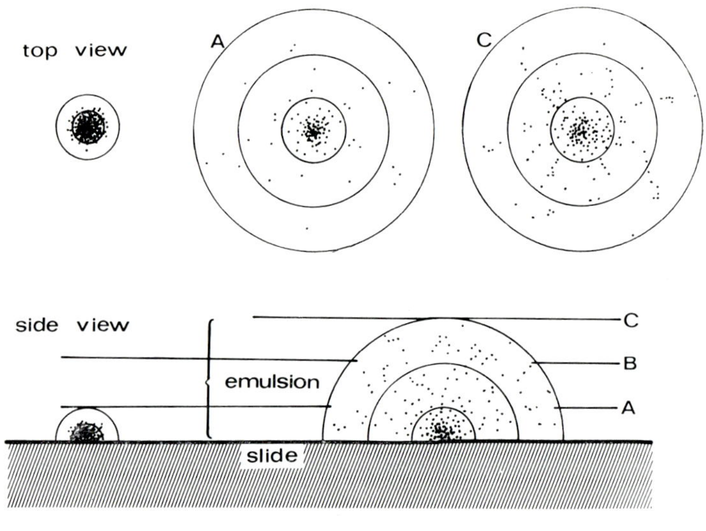

Fig. 5.1. Autoradiographic resolution in relation to the type of isotope and thickness of emulsion calculated from nuclei of *Allium cepa*, labeled with 3-H-TdR and 14-C-TdR, resp. Note the dense, small halo of silver grains formed by a low energy β-emitter (top view, left) and the large halo formed by a medium energy emitter (top view, C). Lines A, B and C represent emulsion layers of 2, 5, and 8 μm, resp. A and C in the top view and side view refer to each other. As described in the text (5.6.3.2), reducing the emulsion thickness in the case of medium or high energy β-emitters, quickly reduces the density of silver grains outside the source to background levels (top view, A). This clearly improves the resolution. Radiating sources are in the centres of circles and hemispheres. The silver grains outside the circles have been excluded.

crystals should be large, facilitating microscopic evaluation. Emulsion types Ilford K5, Eastman-Kodak NTB 3 may be utilized, but in most cases G5 is preferable.

For *quantitative problems on a comparative scale* both types of autoradiographs can be used. In grain density autoradiographs with 14-C, 35-S and 32-P the yield of silver grains depends on emulsion thickness. A constant thickness should therefore always be used, a demand not easily fulfilled with liquid emulsion layers (LEBLOND et al., 1963). Quantitative work with 3-H and 125-I involves the critical problem of β-selfabsorption (5.12.3).

For *determination of the absolute amount* of incorporated precursor track autoradiographs with strong β-emitters are most efficient, because usually a smaller error arises in counting tracks rather than grains. The isotope used should be a strong β-emitter because the percentage of electrons absorbed by the tissue is then minimal (5.12.3). The emulsion must be more sensitive and its layer thicker in proportion to the strength of the β-emitter. Again, one selects emulsions with large silver crystals (e.g. G5).

5.6.4. Preparation of the Autoradiographs

5.6.4.1. Darkroom

The darkroom must contain the appropriate safelight for the individual emulsion types: for Kodak products this is obtained by filters of the Wratten series No. 2 (Kodak), or of Agfa R4 and even R5, Ilford emulsions permit work with higher light intensities, Wratten OC or Ilford "s" No. 902 are recommended. In all cases 15 W bulbs at a distance of about 1 m from the working surface should be used. The darkroom should be dustfree, its temperature about 18—20°, its relative humidity between 40—60% during the stripping of the films (3.4.2) and during the drying of the emulsions (5.12.1).

5.6.4.2. The Liquid Emulsion Procedure

The stripping film procedure has already been reviewed in section 3.4. Here we shall describe only the basic principles of the liquid emulsion technique. For the liquid emulsion technique are recommended: 1 dipping vessel (minimal diameter 3.0 cm, height 6.5 cm for slide sizes of 7.5 × 2.5 cm), two 25 or 50 ml tall beakers, several drying racks or cooled metal or glass plates for the emulsion-covered slides, a gentle cold fan (5.12.1), slides, slide boxes of black plastic for 25 slides (Fig. 5.3), black Sellotape, a towel and Kleenex tissues. For Ilford emulsions in addition a bone or glass spoon and a glass rod.

a) *Preparation of the Emulsions for the Desired Layer Thickness.* The *Eastman-Kodak emulsions*, taken from the refrigerator, are placed under safelight in the waterbath (40—45°) for 30 min and, after melting, are poured into the prewarmed dipping vessel. After 5 min one tests the emulsion by dipping a slide and inspecting the emulsion under safelight for air bubbles. About 10 ml of emulsion is needed for 50—70 slides. Eastman-Kodak emulsions can be melted repeatedly. Undiluted, they give layers of about 5 μm thickness.

Undiluted *Ilford emulsions* will give layers of about 30 μm thickness after dipping. For 4—5 μm layers 15 ml of water or 2% glycerol solution in water (5.12.1) is pipetted into the dipping vessel and is heated to 40 to 45°C. Then the cold gelatine-like Ilford emulsion (e.g. K2) is added with a bone or glass spoon up to a volume of 30 ml (marking ink). The mixture is heated in the water bath for about 10 min and is stirred carefully. After 5 min the emulsion is tested for homogeneity as described above for NTB emulsions. Air bubbles in the emulsion can also be removed by filtering it through muslin or fine cheesecloth fastened with a rubber band over the top of the dipping beaker. If many preparations are to be made, it is advantageous to keep some diluted emulsion in reserve in the water bath.

For *high resolution autoradiographs* (layer thickness 1—2 μm, cf. 5.6.3.2) the emulsions are diluted further. For Ilford emulsions dilutions up to 1 : 8 are possible but dilutions of 1 : 3 to 1 : 4 are usually sufficient for light microscopy. The final concentration of glycerol in the emulsion should always be 1%.

For *track autoradiographs* the viscous Ilford emulsions G5 and K5 are used to advantage, NTB 3 or -2 can also be used. As a rule 30 μm thick layers are sufficient for 14-C and 35-S; for 32-P layers up to 150 μm are necessary (Levi et al., 1963; Guidotti and Setti, 1956). The emulsions are used undiluted or 10 ml of a 4% solution of glycerol and 40—50 ml of Ilford emulsion are mixed thoroughly. For calculation the amount of emulsion necessary for a specified final thickness: 90 μl undiluted Ilford emulsion per cm^2 will give an emulsion thickness of about 100 μm in dry state (shrinking factor upon drying about 1 : 8).

There is no general agreement about repeated melting and duration of storage of Ilford emulsions. The low sensitivity types are often diluted and put into small bottles for storage in a refrigerator. They may be stored 2—6 months depending on the amount of background, which can be tolerated. According to our experience the background increase under these conditions is low, about 3 silver grains per 100 μm^2. However, the emulsion usually contains bacteria which are seen in phase contrast.

b) *Covering the Slide with Emulsion.* For the preparation of layers up to 40 μm, two slides, back to back, are dipped into the properly diluted emulsion. After 2—3 sec the slides are lifted out vertically and if necessary, the surplus emulsion is drained in the vertical plane onto filter paper. The slides can also be dipped separately. In either case one wipes off the emulsion from the back with a moist towel. Thereafter they are best placed horizontally on a precooled plate (5.12.1).

For the production of thicker layers (i.e. more than 10 μm or 40 μm with Eastman-Kodak and Ilford emulsions, resp.) one pours or pipettes the emulsion onto prewarmed slides. In addition, 2 beakers containing demineralized water are placed into the waterbath (Rogers, 1969). One beaker contains graduated pipettes (glass tubes may also be used) of the desired volume with medium sized orifices, the other serves to rinse used pipettes. A warm prewatered pipette equipped with a rubber bulb is used for sucking up the amount of emulsion needed for the desired thickness. Do not suck up the emulsion higher than to the volume mark. The emulsion is run onto the prewarmed slide and distributed with a glass rod or with a soft paint-brush. With the rod or brush any air bubbles can be moved to the edge of the slide.

c) *Drying of the Emulsion on the Slide, and Preparing for Exposure.* The drying of the preparation should begin in a horizontal plane during the gel phase of the emulsion. Caro and v. Tubergen (1962) suggest placing the dipped slides onto cooled plates. The preparations are dried in the dark, in a gentle air current and at a rel. humidity of 40—50%. NTB emulsions and AR-10 need about 25 min, Ilford emulsions are better dried for a somewhat longer time (Rogers, 1969). The drying of thick layers is very critical; if it is carried out too quickly, background developes. 1—2 hrs drying at the conditions described above is usually sufficient.

The techniques just described may be adapted according to the requirements of the double labeling technique (5.7).

Liquid emulsion techniques have also been described in which the slide was immersed and the specimen then brought onto the dried emulsion (mounting procedure, cf. Chapter 3). A combined mounting and dipping method (sandwich technique) has been used by Pickworth et al. (1963) for improving the counting efficiency for 3-H.

d) *Cleaning the Vessels.* The dipping vessels should be cleaned immediately after use, since emulsion residues could lead the background (5.12.1). Foret (1970) suggests enzyme-containing detergents. Rogers (1969) suggests: Hot water, 1 N NaOH for several hours, hot water, acidic treatment, cold water for several hours and demineralized water.

5.7. *Double Labeling Technique*

The differences in the pathlength of β-particles provide the basis for the double labeling technique (5.3). The most widely used label combinations are 3-H/14-C, 3-H/35-S or 3-H/32-P. Double labeling is utilized, when synthesis of different substances is simultaneously studied, or when the percentage of cells synthesizing a substance at a definite time is to be determined in a cell population (e.g. Hilscher and Maurer, 1962; Kiefer and Kiefer, 1969). In the first case two substances differing also in the isotope are applied simultaneously, in the second case the same substance, but labeled with different isotopes, is successively given as precursor.

Usually one works with two film layers, i.e. two liquid emulsion layers. However, combined liquid emulsion/stripping film techniques have also been described (e.g. Kiefer and Kiefer, 1969). The preparation of the autoradiographs does not warrant additional effort, one simply places one layer upon another. Less sensitive emulsions for the lower layer are selected (5.6.3.2) about 2—4 μm thick, with 3-H/32-P even AR-10. These record electrons of low energies. After drying several possibilities exist:

1. The second layer is placed immediately upon the first of the same or of a higher sensitivity. This layer exclusively records high energy β-particles (14-C, 35-S, 32-P). During evaluation the two layers are often difficult to distinguish. Hilscher and Maurer (1962) select emulsions of various grain sizes for this purpose. Inert layers between the two emulsion layers have also been used (5.12.5); Baserga and Nemeroff, 1962).

2. Field et al. (1965) expose, develop and colour the first layer by means of a photographic dye coupling process before mounting the second. This technique facilitates not only the differentiation of the two layers but also circumvents the difficulties which may originate while developing thick emulsions (5.10).

With the aid of double labeling technique, cells labeled with 3-H or 14-C alone can be distinguished clearly. If both isotopes occur together in the same structure, they can only be distinguished from one another if 3-H is present in high concentration relative to 14-C, because the lower layer also records electrons of low energy from 14-C (in 14-C about 15%, Levi et al., 1963, in 32-P a negligible percentage of the total electrons.) The differentiation of 3-H and 14-C label can be improved by choosing suitable proportions of the precursors and exposure times for the different layers.

A second disadvantage can be the loss of resolution in the second layer due to its increased distance from the specimen (5.8).

5.8. *Resolution*

The resolution which can be achieved is important for questions of localization. Resolution can be defined in several ways (DONIACH and PELC, 1950; LAMERTON and HARRISS, 1954; BACHMAN and SALPETER, 1965; CARO and SCHNÖS, 1965). Besides the type of isotope (5.6.3.2, Fig. 5.1) the resolution in autoradiographs depends on several other factors such as quality of fixation, the thickness or type of emulsion (5.6.3.2, Fig. 5.1), the development (5.10), the duration of exposure (5.9; 3.3.1), and especially on the distance between the source of radiation and the emulsion. The resolution rapidly diminishes if the specimen and the emulsion are separated (e.g. by inert layers; 5.12.5, cf. STEVENS, 1950 and Kodak data booklet SC-10) or when thick specimens are used (scattering or absorption of electrons, see 5.12.3).

In practice one usually can attain a resolution with 3-H below 1 μm, with 14-C and 35-S around some 2—4 μm, and with 32-P around 5—8 μm. For 32-P, in many cases resolution can be improved by track autoradiography (5.6.3.2).

The high resolving power and validity of 3-H autoradiographs is shown in Fig. 5.2. From electron microscopic investigations it is known that DNA in plastids and mitochondria clumps in some aldehyde- and OsO_4-fixatives. If labeled the resulting radiating point source would permit autoradiography of high resolution. In this way, HERRMANN (1970) found evidence for several DNA-regions within the chloroplast down to organelle diameters of 2.5 μm (see inset in fig. 5.4 in HERRMANN, 1970). These conclusions have been confirmed by electron microscopic analyses (HERRMANN and KOWALLIK, 1970 b; KOWALLIK and HERRMANN, 1972).

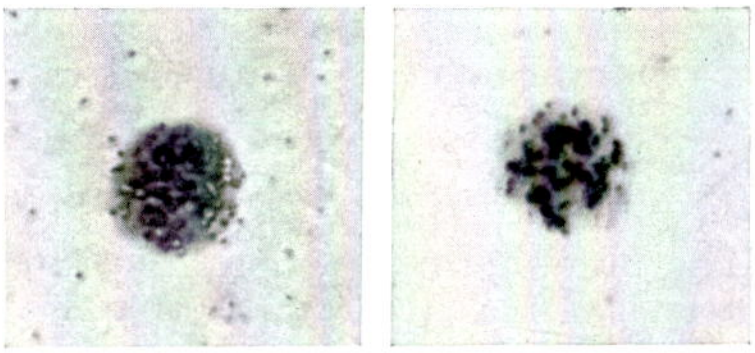

Fig. 5.2. High resolution autoradiography of a chloroplast of *Beta vulgaris*. The same chloroplast is seen in phase contrast (left) and in light field (right). Clusters of grains indicating several regions of DNA are clearly visible in the light field photomicrograph, whereas the organelle can be seen only in phase contrast. Procedure: Samples of *Beta* leaves were incubated for 12 hr in 3-H-TdR (10 μC/ml). The tissue samples were squashed, and the librated chloroplasts were then fixed on the slides with glutaraldehyde, washed in buffer, treated with preheated RNase for 8 hr at 30 °C, covered with AR-10 stripping film, exposed for 5 weeks, and developed in amidole at 14 °C for 20 min (× 1,000; from HERRMANN, 1969)

Fig. 5.3. Storage of the slides in light-tight plastic boxes for exposure. The AR-10 covered slides in the middle of the box are protected from touching the desiccant bags (arrows) by uncovered slides. Left: Opened bag containing desiccant. See also text

5.9. *Exposure*

After drying the emulsion the slides are best put into *black* plastic slide boxes (Fig. 5.3). The box is sealed with Sellotape and the slides are stored in a refrigator (5.12.1). To prevent latent image fading in Kodak emulsions, a bag of desiccant (silica gel or $CaCl_2$) is also placed in the box (Fig. 5.3 arrows; 5.12.1). An uncovered slide is placed between the drying material and the preparations, since the emulsion may burst off the slide on contact with the drying agent.

Duration of exposure and number of silver grains are correlated (cf. below, 5.8 and Chapter 1.3.1). One exposes as briefly as possible; exposure time varies between several hours and several months (5.4.2, Table 5.1). The suitable time is found by developing a series of similar preparations at various intervals (e.g. after 1, 4, 16 etc. days of exposure). The optimal time depends (1) on the type of isotope and on the concentration of incorporated label per unit area, (2) on the type of emulsion used (5.6.3.1) and (3) on the information desired (5.6.3.2; 3.3.1).

To facilitate the extrapolation of tracks, track autoradiographs are exposed only a few days. Emulsions with small crystals need a somewhat longer exposure as compared to emulsions with larger crystals and equal sensitivity. An excessively long exposure diminishes the resolution even in grain density autoradiographs (1.3.1; 5.8). In addition the possibility of latent image fading is increased (5.12.3).

To prevent latent image fading, the relative humidity during exposure should be low to our experience even with Ilford emulsions. However, there are conflicting reports about the conditions of humidity for the exposure of Ilford emulsions (ROGERS, 1969). Concerning temperature see Chapter 5.12.1. When low rates of incorporation are expected, even oxidizing atmosphere should be avoided (5.12.2). In general a CO_2 atmosphere suffices. To obtain this, some dry ice is poured directly into the exposure box before sealing (see also HERZ, 1959).

5.10. *Development and Fixation*

During development of the preparation, the yield of silver grains in the emulsion increases initially. Later it remains constant and finally it increases again, creating background. Optimal development is obtained when the system is stopped at the plateau phase. This depends on the kind of preparation (isotope, emulsion), the kind and concentration of the developer, and also on the temperature and the time of development. Optimal development must be sought for with a series of control preparations, the values listed below can be regarded as a guide. The developers usually used for light microscopic autoradiography are based on hydroquinone, amidol and metol. Amidol developes in the neutral area (pH about 7.2) and therefore quite sparingly (cf. 1.5.3). In contrast to electron microscopic requirements it is in many cases immaterial, which type of developer is used.

Developer 1) Kodak D 19b or D 19 (hydroquinone-metol basis, pH about 10.5) diluted 1 : 1, or undiluted using lower temperatures of development. See also information sheet contained in the packages.

2) Eastman-Kodak Dektol.

3) Amidol: 1.2 gm amidol; 4.5 gm sodium sulfite, anhydrous; 1 ml of 10% KBr solution to 250 ml of deionized water. Keep dark.

After exposure the preparations are transferred to metal or glass slide racks and then into the developing solutions of desired temperature. Development should be performed in darkness and at constant temperature (14—20°) to prevent swelling, extension, production of bubbles or detachment of the emulsion. Gentle agitation speeds up the development and removes air bubbles which could adhere to the emulsion.

Stripping film AR-10: Developer (1), 5 min, 18—20° or developer (3) 20 min, 15°; NTB-emulsions: developer (2) diluted 1 : 1 to 1 : 3, 2—5 min at 15°; Ilford emulsions developer (3) or (1).

For sensitive emulsions as well as for track autoradiographs, high resolution autoradiographs, for stained preparations, and for evaluation in the dark field or in the phase contrast, the developer is diluted (1 : 1 to 1 : 6). For reasons of diffusion, thick layers are preswollen in the developer for 30 min at 4° and are then developed for 15 min. For grain density autoradiographs, which must be evaluated in a comparative and quantitative manner, the developer is used only once. The development temperature and the concentration of the developer should be low so that small variations in development time will be unimportant.

Rinse the slide racks in water for 15—30 sec; for thick layers a stop bath (1% acetic acid, 0 °C) for up to 10 min is recommended. Fix in a 30% solution of sodium thiosulfate or in Kodak Unifix. Fixation time should last twice the time required for clearing the film layer (usually 4—10 min; with thick layers 1—1.5 hrs and several changes of the fixing solution).

Wash in gently running tap water for 20 min, up to 2 hrs with thick layers. If thiosulfate is not removed the silver grains may be dissolved or diminished when acidic treatment (e.g. staining) follows (ROGERS, 1969). The preparations are dried in a gentle dustfree air current.

5.11. *Evaluation*

For analysis of the relation between the silver grains in the emulsion and the various cell structures there are the following possibilities:

1) Analysis in the light field without stain or with staining before or after covering the preparation with emulsion.

2) Analysis with the phase contrast microscope.

3) Analysis in the dark field.

For automatic evaluation see Chapter 1.

5.11.1. Staining prior to Covering the Preparation with Emulsion

It is advisable that dyes do not produce chemography (5.12.2) or be washed out of stained structures during processing. These demands are met by Feulgen or Feulgen-fast-green (KAMINSKI, 1955).

1. After fixation wash tissues or sections 5 min in deionized H_2O.

2. Hydrolyse by anyone of the following three methods: for 12—24 hrs in 10% PCA at 20 °C (see KASTEN, 1965), for 30—80 min in 5 N HCl at 20 °C, or for about 10 min in 1 N HCl at 60 °C.

3. Stain for 2 hrs in leucobasic fuchsin solution in the dark (Leucobasic fuchsin: 500 mg p-rosalinine or "basic fuchsin" is dissolved in 8 ml 1 N HCl, then dilute with 5% solution of K_2SO_3 in deionized H_2O to 80 ml and add 120 ml solution A (see 4). The pH should be adjusted to 2.3. Close the container with a stopper and store the solution in a cool, dark room for 24 hrs. If the solution is then transparent and straw-coloured, it is ready for use).

4. Wash in three changes of sulfite-water, allow 10 min in each change (sulfite-water; solution A: 7.51 g/l glycine and 5.85 g/l NaCl; solution B: 1 g K_2SO_3 is dissolved in 100 ml 0.1 N HCl. 120 ml of solution A is added to 80 ml of solution B. The solution B should be freshly prepared).

5. Wash thoroughly in water.

6. If counterstaining is desired immerse in 2% fast green dissolved in 5% acetic acid for 30 sec.

7. Wash in water and differentiate in 80% ethanol.

8. Immerse for some minutes in 50% ethanol.

9. Wash again in water.

The removal of the cover glass from the squash preparations should be done quickly to prevent the stain washing out; for this CO_2 is blown from a storage tank through a jet (as used for cryostats) onto the back of the slide. Dry-ice sublimes on the slide and the specimen freezes solid. The cover glass may be popped off easily by inserting a razor blade at one corner (CONGER and FAIRCHILD, 1953).

Further stains used were carmine acetic acid and orcein acetic acid. According to STÖCKER and MÜLLER (1967) orcein acetic acid, according to FICQ both dyes initiate chemography, but we could not demonstrate this, when spore mother cells of *Sphaerocarpos* were squashed in orcein- or carmine acetic acid and were dried and covered with AR-10 after removal of cover glasses (ZILCKEN, 1971).

5.11.2 Staining after Photographic Processing

For this purpose, stains should not be used which require hydrolysis such as Feulgen staining (THURSTON and JOFTES, 1963). Other dyes stain the gelatine of the emulsion so strongly as to make light microscopic investigation impossible. This is a disadvantage, especially with thick emulsion layers (THURSTON and JOFTES).

The temperature of the solutions should not exceed 20° while staining. Dry preparations should be soaked for 30—60 min prior to staining. All staining solutions should be filtered before use, to prevent precipitations on the wet emulsion. The following stains have proven to be useful.

1. Hematoxylin (for general morphology); 2. Toluidine blue (for chromosomes); 3. Methyl green — Pyronin (methyl green stains highly polymerized DNA, pyronin stains RNA).

Harri's Hematoxylin: Stain for 5 min in a hematoxylin solution prepared as follows: Dissolve 5 g hematoxylin and 3 g aluminium ammonium sulfate in 1,000 ml 50% ethanol by heating, then add 6 g mercuric oxide (use only the red powder) and boil for 30 min. Filter, then bring up again to the original volume with 50% ethanol. Acidify with 1 drop 1% hydrochloric acid to each 100 ml solution. After staining rinse quickly in tap water and differentiate by dipping 3—5 times in acidic 70% ethanol (1 ml 1 N HCl to 100 ml 70% ethanol).

Ehrlich's Hematoxylin: Stain 30—50 min in the following prepared solution: Dissolve 1 g hematoxylin crystals in 50 ml absolute ethanol and add 50 ml deionized water, 50 ml glycerol, 5 ml glacial acetic acid and aluminium ammonium sulfate (in excess). Ripen the solution in the light, until the solution acquires a dark red color (about 1 month). Differentiate in 0.2% aqueous HCl 30—60 sec after staining blue the autoradiographs in rinsing tap water 30—60 min.

Toluidine blue: Immerse the slides in a 0.5% aqueous solution of toluidine blue (w : v; pH 6.3) for 1—2 min. Wash and differentiate in 95% ethanol for 5—10 min.

Before staining with toluidine blue or after staining with hematoxyline one can use eosin to discolor the cytoplasm. Immerse in a 1% aqueous solution of eosin y for 3—5 min. Wash in water, until the film has lost most of eosin.

Methyl green-Pyronin: Immerse the autoradiographs for 30 min in methyl green-pyronin. [Dissolve 2 g of pyronin y (Fischer Chemical Co.) or pyronin GS (Chroma-Gesellschaft) in 100 ml of deionized H_2O. Add chloroform and shake mixture in a separating funnel until the layer of chloroform becomes colorless. Separate the dissolved dye. Similarily, prepare a 2% solution of methyl green and extract it with chloroform. The solutions can be kept as stock.] For use, mix together 12.5 ml of pyronin solution and 7.5 ml of methyl green solution and add 30 ml of deionized water. Wash in deionized water.

For the use of other dyes see DARLINGTON and LA COUR (1969), RUTHMANN (1970), and THURSTON and JOFTES (1963).

After staining, the preparations are dried in air or dehydrated by treatment with increasing concentrations of ethanol; they may then be embedded under cover glasses in Euparal and Eukitt for storage.

5.11.3. Evaluation by Phase Contrast and Dark Field Microscopy

The evaluation by phase contrast gives satisfying results only when the section thickness does not exeed more than 1 μm, and provided that the silver grains in the layer are small (cf. also 4.4.11). The efficiency of phase contrast microscopy is shown in Figs. 5.2 and 5.5. In the dark field small silver grains are essential for analysis (see Chapter 1).

5.12. *Background, Other Artifacts, Controls, Efficiency, Restripping Technique, Inert Protective Layers*

5.12.1. Background

Every silver grain in the emulsion, which is not produced by radiation from the specimen is designated as *background.* It is always present, even with all precautions taken, and effects the quality of the autoradiographs adversely.

Inadequate safelights or excessive proximity between the safelight and preparations produce a homogeneously distributed silver grain veil (5.6.4.1). Sawicki and coworkers (1968) have determined that for AR-10 temperature-dependent background during the exposure from about —20 to +20°C produces traceable, but small variations within this range. Low temperature should always be used when incorporation rate is low (long exposure) and is recommended when the risk of high or spontaneous background, especially in sensitive emulsions is to be minimized.

One finds high background in overaged emulsions. Over-concentrated developer, high development temperature as well as over-development, trace elements in solutions, isotopes in the gelatine or in the slides (14-C, 40-K), dirty glass vessels, are further reasons for silver grain production. The various types of cosmic rays can only be shown by secondary effects, because in most cases their energy to produce a latent image is too high.

Also, pressure and pull on the emulsion must be avoided. Considerable stress develops during the drying or developing processes of the emulsion due to the great reduction of volume (5.6.4.2). Rapid or too rigorous drying give raise to background, especially with thin or sensitive emulsion layers. In specimen with irregularly shaped surfaces, structure-specific incorporations of the label can be imitated by pulling-stress on the edges.

In botanical objects this can be critical because an extended topography of a specimen can result from differences in volume reduction of the individual cell components during drying. In large-celled algae silver grains have sometimes been found along the cell outlines even in "cold" controls, although they are missing on the cell surface (HERRMANN and KUTZELNIGG, unpubl.; Fig. 5.4).

The relation between drying time and background production have been analysed for the stripping film AR-10 by SAWICKI and PAWINSKA (1965). In Ilford emulsions, 1% glycerol is added to the emulsion to slow the drying process (WALLER, 1961; 5.6.4.2). Cooling is also recommended while drying, especially for thin layers (CARO and v. TUBERGHEN, 1962).

A *correction for background* must be made for grain density autoradiographs. As a rule, several counts of silver grains in equal areas of the same preparation outside the specimen provide an estimate of the background. A rather constant thickness of the emulsion is, however, necessary (cf. also 1.6.5).

Diminuation of background: One can use the conditions which lead to the destruction of latent image to reduce unwanted background prior to exposure. This can be accomplished by exposing the slides for about 1 h at 100% rel. humidity at 37° (WALLER, 1961) or by placing the slides on glass rods over a solution of 1—6% H_2O_2, contained in an airtight box (cf. CARO and v. TUBERGHEN, 1962; MARIMUTHU, 1970).

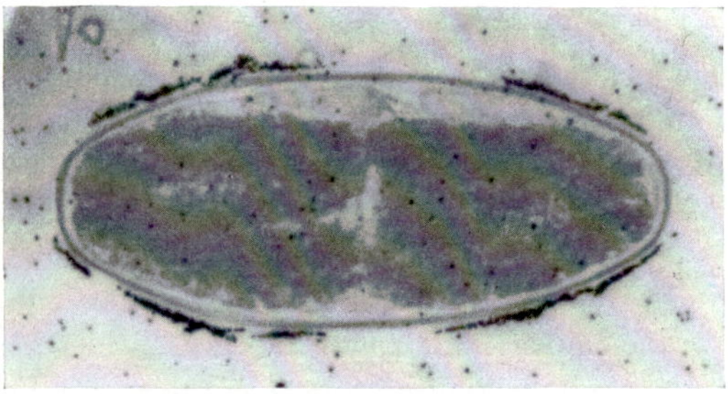

Fig. 5.4. *Netrium digitus* (desmid) focussed such that both alga and silver grains may be seen. Algae were incubated in 3-H-TdR (1 μC/ml) for 6 hr, then fixed in glutaraldehyde. Ilford G5 emulsion (diluted with water) was used. Quick drying (within 8 min) leads to high random background and also to characteristic silver grain distributions along the cell outlines due to pulling stresses. See also text (× 200)

5.12.2. Chemography and Other Artifacts

Compounds diffusing from the specimen into the emulsion by direct contact can reduce Ag^+ of the emulsion or can have oxidising effects (cf. 5.12.1, or DEUCHAR, 1962). These special cases of background are designated as chemography; in the first case chemography is positive, in the second it is negative. Chemography may produce characteristic silver grain distributions above the specimen which can lead to misinterpretations. Chemography can be reduced or omitted by exposing the completely dry preparation in a freezer or by using inert, thin and impermeable layers between emulsion and specimen (5.12.5). For proof of chemography see 5.12.3.

In this connection it should be pointed out that absence of silver grains can have other reasons than negative chemography. None or little uptake of the precursor by the cell could have occurred, the concentration of precursor in the incubation medium could have been low, a pool of cold precursor may have existed in the cell, the application of the precursor may occur outside the S-phases, or the exposure may have been too short. In all these cases, the silver grains of the background should be visible, especially at the cut edges of the stripping film. In experiments using 3-H, the total self-absorption of the electrons by the tissue has to be considered if the specimen is too thick and the isotope distributed unfavourably (5.12.3).

"Insoluble" cell components may also not necessarily be resistant to dissolution during every step to treatment, e.g. after enzyme digestion. Partial loss of incorporated label may occur chemically (partial hydrolysis during Feulgen treatment, change from non-polar to polar washing fluids and *vice versa*; change from acidic to basic conditions) or physically, by structural changes prior to fixation (e.g. fig. 6 in HERRMANN, 1970). Such a loss during the preparation can often be recognized by liquid scintillation counting of solutions from the various stages of treatment.

It is evident, therefore, that no conclusions can be drawn from simply the presence or absence of silver grains in an autoradiograph. *Each* alternative conclusion must be excluded by proper controls before one can form a *valid* conclusion. Generally the scheme of controls outlined in the next section can be used to guard against misinterpretations of the data.

5.12.3. Controls, Efficiency

First one must determine, whether an autoradiographic result is based on chemography (1) or on incorporated label (2). In the latter case, one must then determine which compounds were labeled by the precursor. For judgements of specificity and type of incorporation the selection of the precursor alone is not sufficient! Additional controls are required for quantitative autoradiography (3).

(1) a) Positive chemography is determined by exposure of a "cold" preparation, i.e. one which was not treated with radioactive precursor.

b) Negative chemography and bad development are determined in a "cold" preparation which has been placed for 30 sec near a safe light before exposure. Often negative chemography can also be recognized by differences in background between the specimen and its surrounding area.

Both types of preparations can also give indications if the conditions of the autoradiography have not been optimal (e.g., if latent image fading occurs; see (3) below and 5.9).

(2) Specific incorporation may be obfuscated by several possibilities.

a) Attachment of precursor to "insoluble" compounds during incubation and fixation may occur. We have already mentioned this possibility in the case of amino acids during fixation (5.5). There is also some evidence for enzymes that attach precursors onto structures (e.g. the probable linkage of ATP to t-RNA; SUYAMA and EYER, 1968).

b) The possibility that during preparation radioactive label is adsorbed onto cellular structures must also be excluded. This is especially important for squash preparations with unfixed material (Fig. 5.5).

c) Specific incorporation may have occurred but through a different process, e.g. DNA repair synthesis rather than DNA replication.

d) It must be demonstrated, especially when working with high doses of precursor that no unwanted metabolism occurred during incubation. After application of 3-H-TdR to spore mother cells of *Sphaerocarpos*, one finds label which appears to be distributed over the entire cell (in contrast to the case of desmids shown above; Fig. 5.6). Here the likelihood of unwanted metabolism was great (cf. also 6.12.2). However, squashing showed that the label was localized in the cytoplasm, presumably in mitochondria or proplastids (Fig. 5.6).

e) Finally, isotope exchange of 3-H labeled precursors with surrounding structures must also be excluded (5.4.1).

Fig. 5.5. One test for specificity of DNA labeling in chloroplasts of *Beta vulgaris*. Adsorption artifacts must be excluded when fixation is performed after squashing of the tissue samples. Left: light field autoradiographs; right: phase contrast autoradiographs of the same preparation. From top to bottom; top: chloroplasts labeled, cyto- or nucleoplasm unlabeled. Middle and inset at bottom: chloroplasts unlabeled (arrows), cyto- or nucleoplasm labeled. Bottom: chloroplasts labeled, cyto- and nucleoplasm labeled. N = squashed nucleus. Procedure: tissues incubated 12 hr in 3-H-TdR as in Fig. 5.3. Tissue samples are strongly squashed to produce thick suspensions of chloroplasts and smears of cytoplasm. Fixed in glutaraldehyde, washed in 0.2 N PCA and treated with preheated RNase. Autoradiographic conditions as in Fig. 5.2. (× 1,200; from HERRMANN, 1970)

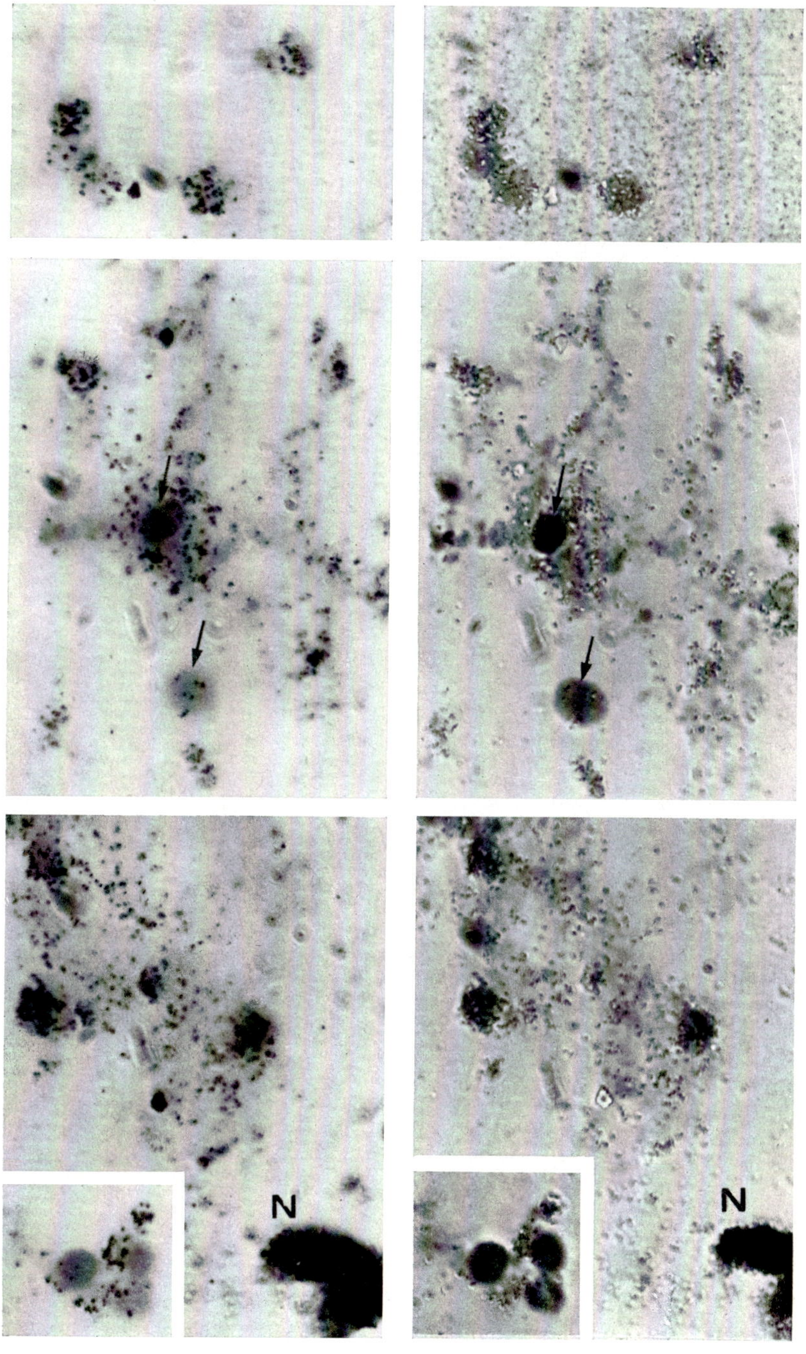
N
N

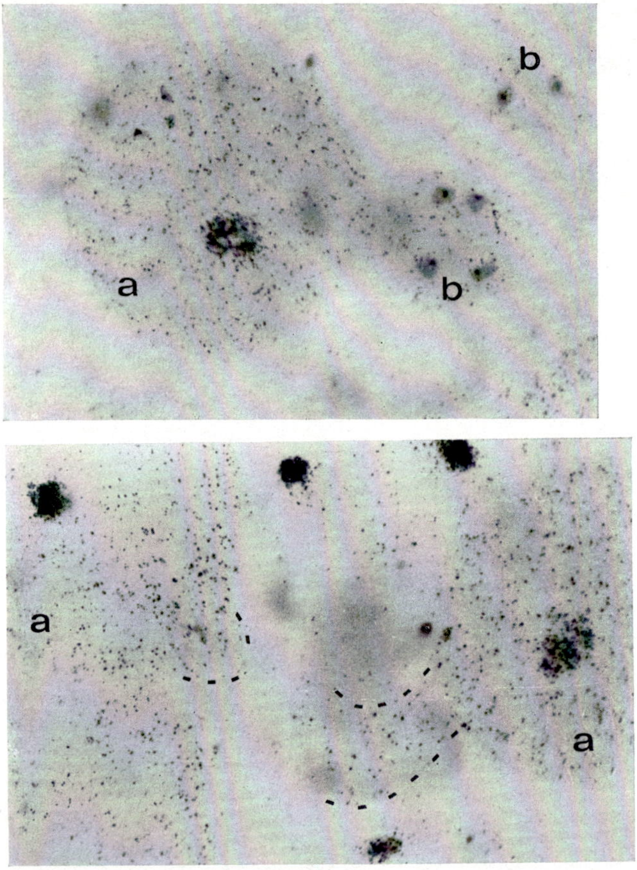

Fig. 5.6. Analysis of the premeiotic DNA S-phase in the liverwort *Sphaerocarpos donnellii* (from ZILCKEN 1971). Upper: (a) Spore mother cells with labeled nuclei; label also is distributed over the entire cell surface, (b) multinucleate tapetal cells, nuclei unlabeled. Lower: Nuclei of spore mother cells and tapetal cells labeled. The extrusion of cytoplasm (dotted lines) indicates labeling of cytoplasm rather than of cell surfaces. See also text. Procedure: Tissue pieces containing sporogones were incubated for 12 hr in 3-H-TdR (10 μC/ml in White medium containing 10^{-4} M FUdR and 10^{-5} M uridine). The sporogones were fixed in Carnoy's fluid II, postfixed in Bouin-Allen fixative, stained with Feulgen, squashed on the slides, covered with AR-10 stripping film, exposed for 10 days and developed in Kodak D19. ($\times$ 650)

Specificity and often type of incorporation can be tested by enzymatic treatment, by differential extraction with acidic or basic solutions, or by treatment with specific inhibitors. However, some of the above mentioned kinds of artifacts are difficult to exclude by autoradiography alone.

After fixation, tissues, squash preparations, or sections are treated with enzymes before covering them with emulsion (DNase: 50 μg to 1 mg/ml in 0.05 M phosphate buffer, pH 7.2, containing 6 mM $MgCl_2$ for up to 20 hrs at 20—40 °C; preheated RNase A: 20—500 μg/ml in phosphate buffer or any other buffer 0.05 M, pH about 7.0, trypsin and Pronase according to KOWALLIK and HERRMANN, 1970a). Then one observes which enzyme causes the diminuation or disappearance of label. However, final conclusions should be drawn only after a positive reaction is observed. Steric hindrance of digestion by fixation or problems due to permeability of the enzymes could lead to misinterpretations (cf. difficulties in digestion of DNA in fixed chloroplasts with DNase by SHEPARD, 1965; SPREY, 1968 and HERRMANN and KOWALLIK, 1970a).

Instead of enzymatic treatment less specific differential extraction methods may be used. RNA can be selectively removed by a 10 min treatment with 5% TCA at 60°, or 1 N HCl for 10 min at 60° (Feulgen), or 10% PCA for 12—16 hrs at 4° (KASTEN, 1965). All nucleic acids are destroyed by 5% TCA or PCA at 90° in about 15 min.

The specificity of incorporation can also be tested by treating the specimen with specific metabolic inhibitors. These antimetabolites can be added to the incubation medium simultaneously with the precursor. In most cases it may be of advantage to preincubate specimen with inhibitors. Protein synthesis occuring on the 80 S ribosome can be inhibited by cycloheximide, and puromycin, chloramphenicol or rifamycins inhibit protein synthesis on the 70 S ribosome. DNA-dependent RNA synthesis is inhibited by actinomycin D and DNA synthesis by mitomycin C. Most biological systems respond to these drugs over a range of 0.1—100 μg/ml.

If only a small proportion of the cells in a preparation has been labeled, treatment immediately after fixation will not always yield sufficient information. One can either mount serial sections of a sample onto different slides and treat them in various ways, or one can restrip a preparation. If no special precautions have been taken when restripping, it must be determined if the first development and fixation has produced basic changes. The film is removed from the preparation, which is then treated enzymatically and covered with fresh emulsion (5.12.4).

(3) Quantitative evaluation requires additional controls, including especially the determination of loss of label during the preparation (5.12.2), latent image fading (tested as discussed above, (1) b), and efficiency of counting. Quantitative autoradiographic work on a comparative scale (by comparison of visually determined numbers of silver grains (cf. 3.3.1) or tracks) can be done without much difficulty; absolute measurements of radioactivity require a greater effort (ROGERS, 1969). For automatic measurements of silver grain blackening, see Chapter 1.

The efficiency of counting radioactivity can be defined absolutely as the percentage of decay events in a sample compared to those that are recorded by the emulsion. Alternatively the relative efficiencies of various emulsions can be compared.

Both the efficiency and the resolution (see again 5.8) depend upon the type of isotope used, the thickness and density of the specimen, the type and treatment of the emulsion, the conditions of exposure (5.9, STUBBLEFIELD, 1965), and the distance between specimen and emulsion (5.8).

The problem of self-absorption: As has been shown, by MAURER and PRIMBSCH (1964) for example, the yield of silver grains follows a saturation curve with increasing thickness of the specimen. The plateau is reached when the thickness of the specimen equals the track radius of the isotope. The lower the mean energy of electrons of a given isotope and the denser the structure of the specimen, the more readily this plateau is reached. Only in the early linear portion of the curve are grain yield and the absolute amount of radioactivity proportional.

As is to be expected, the thickness and density of a specimen are of great importance in quantitative autoradiography with 3-H and 125-I. MAURER and PRIMBSCH (1964) have determined the linear range of the curves for nucleoli, caryoplasm, and cytoplasm of liver cells with dry masses of 95, 18 and 25 $\mu g/cm^2$ respectively to be 0.1, 0.5 and 0.5 μm resp. The saturation thicknesses of these structures were 1 μm, 2—3 μm and 1 μm respectively (cf. also PERRY et al., 1961; FALK and KING, 1963). HENDLER (1959) gave counting efficiencies of 14-C for various thicknesses of sections and showed that sections of 1—2 μm produced satisfactory efficiency. Efficiency up to 100% can be obtained with 32-P (LEVINTHAL and THOMAS, 1957; CARO and SCHNÖS, 1965). For sections up to 7 μm the fraction of emitted β-particles absorbed by the tissue is negligible.

The influence of the type of emulsion on the grain yield has been studied by RON and PRESCOTT (1970) in 7 different types of emulsions. The authors have estimated 1 (for K 2) to 15 (for NTB 3) silver grains per 100 disintegrations of 3-H (but see also KISIELESKI et al., 1961; MAURER and PRIMBSCH, 1964). Under comparable conditions one expects about 5—10 times this grain yield for 14-C and 32-P. The emulsion must be dry to be fully sensitive (RON and PRESCOTT).

5.12.4. Restripping Technique

For controls or for quantitative autoradiography (STUBBLEFIELD, 1965) the emulsion can be removed and, if the label has not been destroyed after treatment, the preparation can be covered again with emulsion. The same procedure is applicable for any autoradiograph. Sometimes it is of advantage to place an inert layer between the tissue and the emulsion (5.12.5).

The emulsion layer may be removed chemically (according to Rogers (1969) using 1% KOH at room temperature) or mechanically. The emulsion is then cut either at the edges of the slide and peeled away in a dry state before or after developing (Andresen et al., 1952), or peeled away after swelling in hot water. In the very first case the specimen remains in the same condition for the next autoradiograph.

In spite of careful removal of the layer, silver grains often adhere to the specimen or to the inert layer. They may be removed by treatment with 5% potassium ferricyanide for 40 min, followed by treatment with 25—30% sodium thiosulfate (3 changes of 5 min each) (Rogers, 1969).

5.12.5. Inert Protective Layers

Protective layers between specimen and emulsion are employed to avoid the effects of chemography (5.12.2), between both emulsion layers in double labeling experiments (5.7), and in the restripping technique (5.12.4).

Protective layers should be inert, impermeable and as thin as possible (0.025—0.2 μm). They cannot be used for quantitative work with 3-H (5.12.3) or for high resolution autoradiographs (5.8). One great disadvantage of such inert layers can be the bad adherance of the emulsion to the slide, especially when the layer is hydrophobic.

Often collodion (dinitrocellulose), formvar (polyvinylformal), PVC and nylon films are used as protective layers. Collodion is used in a 2—3% solution in butylacetate or amylacetate or in ether-ethanol mixtures. Formvar is dissolved in about 0.3% solution of ethylene dichloride or chloroform and PVC is prepared according to Sawicki and Darzynkiewicz (1964). Chapman-Andresen (1953) has tested other compounds, among which is nylon.

The protective layer is mounted either by dipping the slide into the solution, or by applying the solution dropwise. In the latter case, the solution usually spreads quickly and shows interference colours until the solvent has evaporated. In some cases, it is possible to place a drop of the solution onto water and the layer formed can be brought to the slide similarly as for the stripping technique (3.3). Layers of defined thickness can be produced by varying the formvar concentration and dipping time according to the following procedure: Formvar is dissolved in anhydrous chloroform. Slides are treated with curd soap and washed. Traces of soap on the slide will facilitate floating of the layer from the slide. The slide is dipped into the formvar solution and dried quickly with a lamp. Pieces of the desired size are cut with a needle. By moving the slide parallel to the water surface and by blowing gently on the slide the layer will float onto the water surface. However, the layer and slide must not be submerged.

References

ALBERTINI, L.: Étude autoradiographique des synthèses d'acide ribonucléique (RNA) au cours de la microsporogenèse chez le *Rhoeo discolor* (HANCE). C. R. Acad. Sci. (Paris) Ser. D **260**, 651—653 (1965).

ALBERTINI, L.: Étude autoradiographique des synthèses de proteines au cours de la microsporogenèse chez le *Rhoe discolor* (HANCE). C. R. Acad. Sci. (Paris) Ser. D **264**, 2773—2776 (1967).

ALLFREY, V., POGO, B. G. T., LITTAU, V. S., GERSHEY, E. L., MIRSKY, A. E.: Histone acetylation in insect chromosomes. Science **159**, 314—316 (1967).

ANDRESEN, N., CHAPMAN-ANDRESEN, C., HOLTER, H.: Autoradiographic studies on the amoeba *Chaos chaos* with ^{14}C. C. R. Lab. Carlsberg, Sèr. chim. **28**, 189—220 (1952).

ARRIGHI, F. E., HSU, T. C., SAUNDERS, P., SAUNDERS, G. F.: Localisation of repetitive DNA in the chromosomes of *Microtus agrestis* by means of in situ hybridization. Chromosoma (Berl.) **32**, 224—236 (1970).

BACHMANN, L., SALPETER, M. M.: Autoradiography with the electron microscope; a quantitative evaluation. Lab. Invest. **14**, 1041—1053 (1965).

BARNARD, E. A.: Location and measurement of enzymes in single cells by isotopic methods. Part I. Principles and light microscopic application. Int. Rev. Cytol. **29**, 213—243 (1970).

BARNARD, E. A., OSTROWSKI, K.: Acetylcholinesterase. II. Studies on some properties of acetylcholinesterase in its sites at the motor end-plate. Exp. Cell Res. **36**, 28—42 (1964).

BASERGA, R.: Autoradiographic methods. In H. BUSCH (ed.), Methods in cancer research, Vol. I, 45—116. New York, London: Academic Press 1967.

BASERGA, R., NEMEROFF, K.: Factors which effect efficiency of autoradiography with tritiated thymidine. Stain Technol. **37**, 21—26 (1962).

BÉLANGER, L. F., LEBLOND, C. P.: A method for locating radioactive elements in tissues by covering histological sections with a photographic emulsion. Endocrinology **39**, 8—13 (1946).

BERLOWITZ, L., PALOTTA, D., SILBEY, C. H.: Chromatin and histones: Binding of tritiated actinomycin D to heterochromatin in mealy bugs. Science **164**, 1527—1529 (1969).

BIRNSTIEL, M., HYDE, B. B.: Protein synthesis by isolated pea nucleoli. J. Cell Biol. **18**, 41—50 (1963).

BLOCH, D. P., MACQUIGG, R. A., BRACK, S. D., WU, J. R.: The synthesis of deoxyribonucleic acid and histone in the onion root meristem. J. Cell Biol. **33**, 451—467 (1967).

BÖHNER, H.: Untersuchungen über das Wachstum und den Feinbau der Zellwände in der *Avena*-Koleoptile. Planta **50**, 461—497 (1958).

BRYANT, T. R.: Hydrolysis in Feulgen autoradiography. I. Loss of label from S-period nuclei and its subsequent association with non-S-chromatin. Exp. Cell. Res. **56**, 127—133 (1969).

BUONGIORNO-NARDELLI, M., AMALDI, F.: Autoradiographic detection of molecular hybrids between r-RNA and DNA in tissue sections. Nature **225**, 946—948 (1970).

BURKHOLDER, G. D., MUKHERJEE, B. B.: Replication of chromosome fibres in isolated metaphase chromosomes. Exp. Cell Res. **64**, 470—473 (1971).

CARO, L. G., SCHNÖS, M.: Tritium and phosphorus-32 in high-resolution autoradiography. Science **149**, 60—62 (1965).

CARO, L. G., TUBERGEN, R. T. VAN: High-resolution autoradiography. I. Methods. J. Cell Biol. **15**, 173—188 (1962).

CAVE, MC. D.: Incorporation of tritium-labelled thymidine andlysine into chromosomes of cultured human leucocytes. J. Cell Biol. **29**, 209—222 (1966).

CAVE, MC. D.: Chromosome replication and synthesis of non-histone proteins in giant polytene chromosomes. Chromosoma (Berl.) **25**, 392—401 (1968).

CHAKRABATORY, K. P., LORING, H. S.: Incorporation of uracil-2-C^{14} into the nucleic acids of *Neurospora crassa*. J. biol. Chem. **235**, 2122—2126 (1960).

CHAPMAN-ANDRESEN, C.: Thin films for application between emulsion and isotope-containing specimen in autoradiography. C. R. Lab. Carlsberg, Sér. chim. **28**, 529—540 (1953).

CHOUINARD, L. A., LEBLOND, C. P.: Sites of protein synthesis in nucleoli of root meristematic cells of *Allium cepa* as shown by radioautography with (3H)arginine. J. Cell Sci. **2**, 473—480 (1967).

CHURCH, K., WIMBER, D. E.: Meiosis in *Ornithogalum virens* (Liliaceae): Meiotic timing and segregation of H^3-thymidine labelled chromosomes. Canad. J. Genet. Cytol. **11**, 573—581 (1969).

CLEAVER, J. E.: Thymidine metabolism and cell kinetics. North Holland Publishing Co. Amsterdam 1967.

CLOWES, F. A. L.: Protein synthesis in root meristems. J. exptl. Botany **9**, 229—238 (1958).

COHEN, S. S., FLAKS, J. G., BARNER, H. D., LOEB, M. R., LICHTENSTEIN, J.: The mode of action of 5-fluorouracil and its derivatives. Proc. Nat. Acad. Sci. (Wash.) **44**, 1004—1012 (1958).

COMINGS, D. E.: Isolabelling and chromosome strandedness. Nature (New Biology) **229**, 24—25 (1971).

CONGER, A. D., FAIRCHILD, L. M.: A quick freeze method for making smear slides permanent. Stain Techn. **28**, 281—283 (1953).

COOK, J. R.: The synthesis of cytoplasmic DNA in synchronized *Euglena*. J. Cell. Biol. **29**, 369—373 (1966).

COPE, G. H., WILLIAMS, M. A.: Quantitative studies on neutral lipid preservation in electron microscopy. J. R. Microscop. Soc. **88**, 259—277 (1968).

DARLINGTON, C. D., LA COUR, L. F.: The handling of chromosomes. 5th ed. London: Allen & Unwin 1969.

DAS, N. K.: Inactivation of the nucleolar apparatus during meiotic prophase in corn anthers. Exp. Cell Res. **40**, 360—364 (1965).

DE, D. N.: Autoradiographic studies of nucleoprotein metabolism during the division cycle. The Nucleus **4**, 1—24 (1961).

DERMER, G. B.: An autoradiographic and biochemical study of oleic acid absorption by intestinal slices including determination of lipid loss during preparation for electron microscopy. J. Ultrastruct. Res. **22**, 312—325 (1968).

DEUCHAR, E. M.: Staining sections before autoradiographic exposure: Excessive background graining caused by celestin blue. Stain Techn. **37**, 324 (1962).

DONIACH, J., PELC, S. R.: Autoradiographic technique. Brit. J. Radiol. **23**, 184—192 (1950).

DOUGLAS, L. T., KROES, H. W.: Meiosis IV: A reinterpretation of tritium thymidine labelling of meiotic chromosomes, supporting the chiasmatype hypothesis. Genetica **40**, 503—507 (1969).

EBSTEIN, B. S.: The distribution of DNA within the nucleoli of the amphibian oocyte as demonstrated by tritiated actinomycin D radioautography. J. Cell Sci. **5**, 27—44 (1969).

FALK, G. J., KING, R. C.: Radioautographic efficiency for tritium as a function of section thickness. Radiat. Res. **20**, 466—479 (1963).

FEDER, N., O'BRIEN, T. P.: Plant microtechnique: some principles and new methods. Amer J. Botany **55**, 123—142 (1968).

FIELD, E. O., DAWSON, K. B., GIBBS, J. E.: Autoradiographic differentiation of tritium and another β-emitter by a combined colour coupling and double stripping-film technique. Stain Techn. **40**, 295—300 (1965).

FINK, R. M., FINK, K.: Relative retention of H^3 and C^{14} of nucleotides incorporated into nucleic acids of Neurospora. J. Biol. Chem. **237**, 2889—2891 (1962).

FORET, J. E.: Enzyme-containing detergent for removing extraneous emulsion from autoradiographic slides. Stain Techn. **45**, 237 (1970).

FREUDENBERG, K., REZNIK, H., FUCHS, W., REICHERT, M.: Untersuchungen über die Entstehung des Lignins und des Holzes. Naturwissenschaften **42**, 29—35 (1955).

FRIEDBERG, S. H., DAVIDSON, D.: Duration of S-phase and cell cycles in diploid and tetraploid cells of mixoploid meristems. Exp. Cell Res. **61**, 216—218 (1970).

FRIEDKIN, M., TILSON, D., ROBERT, D.: In vitro studies of desoxyribonucleic acid synthesis with C^{14} thymidine. Fed. Proc. **13**, 214 (1954).

GOLDBERG, J., OHAD, J.: Biogenesis of chloroplasts membranes. V. A radioautographic study of membrane growth in a mutant of *Chlamydomonas reinhardi* y-1. J. Cell Biol. **44**, 572—591 (1970).

GOROVSKY, M., WOODARD, J.: Histone content of chromosomal loci active and inactive in RNA synthesis. J. Cell Biol. **33**, 723—728 (1967).

GREEN, M. R., BARR, G. F.: Incorporation of 3H-thymidine into mitochondria. Localization of autoradiographic label to individual mitochondria. J. Histochem. Cytochem. **18**, 354—360 (1970).

GUIDOTTI, G., SETTI, R. L.: Autoradiography of tracks from beta-particle emitters in tissues. Stain Techn. **31**, 57—65 (1956).

HENDLER, R. W.: Self-absorption correction for carbon-14. A new treatment yields a correction factor that is linearly related to the thickness of the sample. Science **130**, 772—777 (1959).

HERRMANN, R. G.: Are chloroplasts polyploid? Exptl. Cell Res. **55**, 414—416 (1969).

HERRMANN, R. G.: Multiple amounts of DNA related to the size of chloroplasts. I. An autoradiographic study. Planta **90**, 80—96 (1970).

HERRMANN, R. G., KOWALLIK, K. V.: Selective presentation of DNA-regions and membranes in chloroplasts and mitochondria. J. Cell Biol. **45**, 198—202 (1970a).

HERRMANN, R. G., KOWALLIK, K. V.: Multiple amounts of DNA related to the size of chloroplasts. II. Comparison of electron-microscopic and autoradiographic data. Protoplasma **69**, 365—372 (1970b).

HERZ, R. H.: Methods to improve the performance of stripping emulsions. Lab. Invest. **8**, 71—81 (1959).

HILSCHER, W., MAURER, W.: Autoradiographische Bestimmung der Dauer der DNS-Verdopplung und ihres zeitlichen Verlaufs bei Spermatogonien der Ratte durch Doppelmarkierung mit C-14- und H-3-Thymidin. Naturwissenschaften **49**, 352 (1962).

HOWARD, A., PELC, S. R.: Nuclear incorporation of ^{32}P as demonstrated by autoradiographs. Exp. Cell Res. **2**, 178—187 (1951).

ITO, M., STERN, H.: Studies of meiosis in vitro. I. In vitro culture of meiotic cells. Develop. Biol. **16**, 36—53 (1967).

ITO, M., HOTTA, Y., STERN, H.: Studies of meiosis in vitro. II. Effect of inhibiting DNA-synthesis during meiotic prophase on chromosome structure and behaviour. Develop. Biol. **6**, 54—77 (1967).

JENSEN, W. A.: The incorporation of ^{14}C-adenine and ^{14}C-phenylalanine by developing root tip cell. Proc. Nat. Acad. Sci. (Wash.) **43**, 1038—1046 (1957).

JENSEN, W. A.: Botanical histochemistry. 408 p. San Francisco, London: Freeman & Co. 1962.

JOHN, H. A., BIRNSTIEL, M. L., JONES, K. W.: RNA-DNA hybrids at the cytological level. Nature **223**, 582—587 (1969).

Johnson, B. F.: Autoradiographic analysis of regional cell wall growth of yeasts. Exp. Cell Res. **39**, 613—624 (1965).

Johnson, B. F., Gibson, E. J.: Autoradiographic analysis of regional cell wall growth of yeast. II. *Pichi farinosa.* Exp. Cell Res. **41**, 297—306 (1966a).

Johnson, B. F., Gibson, E. J.: Autoradiographic analysis of regional cell wall growth of yeast. III. *Saccharomyces cerevisiae.* Exp. Cell Res. **41**, 580—591 (1966b).

Jones, G. H., Craig-Cameron, T.: Crossover analysis by tritium autoradiography. Nature **223**, 946—947 (1969).

Kaminski, E. J.: Histologic processing in autoradiography: Loss of radioactivity. Stain Techn. **30**, 139—145 (1955).

Kasten, F. H.: Loss of RNA and protein and changes in DNA during a 30-hour cold perchloric acid extraction of cultured cells. Stain Techn. **40**, 127—135 (1965).

Kiefer, R., Kiefer, G.: Vergleich autoradiographischer Doppelmarkierung mit mikrophotometrischer DNS-Messung. Naturwissenschaften **56**, 420 (1969).

Kisieleski, W. E., Baserga, R., Vaupotic, J.: The correlation of autoradiographic grain counts and tritium concentration in tissue sections containing tritiated thymidine. Radiat. Res. **15**, 341—348 (1961).

Knox, R. B.: Freeze sectioning of plant tissue. Stain Techn. **45**, 265—272 (1970).

Kopriwa, B. M., Leblond, C. P.: Improvements in the coating technique of radioautography. J. Histochem. Cytochem. **10**, 269—284 (1962).

Kowallik, K. V., Herrmann, R. G.: Variable amounts of DNA related to the size of chloroplasts. IV. Three-dimensional arrangement of DNA in fully differentiated chloroplasts of Beta vulgaris L. J. Cell Sci. **11**, 357—377.

Lamerton, L. F., Harris, E. B.: Resolution and sensitivity considerations in autoradiography. J. Photogr. Sci. **2**, 135—144 (1954).

Lang, W., Maurer, W.: Zur Verwendbarkeit von Feulgen-gefärbten Schnitten für die quantitative Autoradiographie mit markiertem Thymidin. Exp. Cell Res. **39**, 1—9 (1965).

Leblond, C. P., Kopriwa, B. M., Messier, B.: Radioautography as a histochemical tool. In R. Wegmann (ed.), Histochemistry and Cytochemistry, p. 1—31. London: Pergamon 1963.

Levi, H., Rogers, A. W., Weis Bentzon, M., Nielsen, A.: On the quantitative evaluation of autoradiograms. Kong. danske Vidensk. Selsk. mat.-fysiske Medd. **33**, No. 11, 1—51 (1963).

Levinthal, C., Thomas, C. A.: Molecular autoradiography: the β-ray counting from single virus particles and DNA molecules in nuclear emulsions. Biochim. Biophys. Acta **23**, 453—465 (1957).

Liao, S.-H., Hamilton, R. H.: Intracellular localization of growth hormones in plants. Science **151**, 822—824 (1966).

Lillie, R. D.: Histopathologic technique and practical histochemistry. 3. Ed. The Blakiston Division; New York: McGraw-Hill Book Comp., 1965.

Marimuthu, K. M.: Labelling nuclei of progressively developing pollen mother cells and pollen with tritiated thymidine in *Tradescantia.* Stain Techn. **45**, 105—108 (1970).

Mascarenhas, J. P.: Pollen tube growth and ribonucleic acid synthesis by vegetative and generative nuclei of *Tradescantia.* Amer. J. Bot. **53**, 563—569 (1966).

Mattingly, A.: Nuclear protein synthesis in *Vicia faba.* Exp. Cell Res. **29**, 314—326 (1963).

Maurer, W., Primbsch, E.: Größe der β-Selbstabsorption bei der 3H-Autoradiographie. Exp. Cell Res. **33**, 8—18 (1964).

Moens, P. B.: Segregation of tritium-labeled DNA at meiosis in *Chorthippus.* Chromosoma **19**, 277—285 (1966).

Nagl, W.: Correlation of structure and RNA-synthesis in the nucleolus-organizing polytene chromosomes of *Phaseolus vulgaris.* Chromosoma **28**, 85—92 (1969).

Nagl, W.: Rifamycin SV: Effect of RNA synthesis in meristematic cells of *Allium cepa*. Naturwissenschaften **57**, 458 (1970).

Nelson, J. S. R.: Pattern of segregation of labelled DNA at the chromatid level of chromosomes with diffuse centromeres. Canad. J. Genet. Cytol. **11**, 928—936 (1969).

Ostrowski, K., Barnard, E. A., Darzynkiewicz, Z., Rymaszewska, D.: Autoradiographic methods in enzyme cytochemistry. III. Measurements on esterases in the cells of mouse kidney. Exp. Cell Res. **36**, 43—52 (1964).

Ostrowski, K., Barnard, E. A., Sawicki, W., Chorzelski, T., Langner, A., Mikulski, A.: Autoradiographic detection of antigens in cells using tritium-labeled antibodies. J. Histochem. Cytochem. **18**, 490—497 (1970).

Pardue, M. L., Gall, J. G.: Chromosomal localization of mouse satellite DNA. Science **168**, 1356—1358 (1970).

Pate, J. S., O'Brien, T. P.: Microautoradiographic study of the incorporation of labelled amino acids into insoluble compounds of the shoot of a higher plant. Planta **78**, 60—71 (1968).

Peacock, W. J.: Chromosome duplication and structure as determined by autoradiography. Proc. Nat. Acad. Sci. (Wash.) **49**, 793—801 (1963).

Pelc, S. R.: The stripping-film technique of autoradiography. Int. Journ. appl. Radiat. Isotopes **1**, 172—177 (1956).

Pelc, S. R., Howard, A.: Metabolic activity of salivary gland chromosomes in diptera. Exp. Cell Res. **10**, 549—552 (1956).

Perry, R., Errera, M., Hell, A., Durwald, H.: Kinetics of nucleotide incorporation into nuclear and cytoplasmic RNA. J. Biophys. Biochem. Cytol. **11**, 1—13 (1961).

Peters, T., Jr., Ashey, C. A.: An artefact in autoradiography due to binding of free amino acids to tissues by fixatives. J. Cell Biol. **33**, 53—60 (1967).

Pickworth, J. W., Cotton, K., Skyring, A. P.: Double emulsion autoradiography for identifying tritium-labelled cells in sections. Stain Technol. **38**, 237—244 (1963).

Powell, J. B.: Karyological leaf squashes in grasses aided by pectinase digestion at 45 °C. Stain Technol. **43**, 135—138 (1968).

Prescott, D. M.: The synthesis of total macronuclear protein, histone, and DNA during the cell cycle in *Euplotes eurystomus*. J. Cell Biol. **31**, 1—9 (1966).

Quastler H., Sherman, F. G.: Cell population kinetics in the intestinal epithelium of the mouse. Exp. Cell Res. **17**, 420—438 (1959).

Reichard, P., Estborn, B.: Utilization of desoxyribosides in the synthesis of polynucleotides. J. Biol. Chem. **188**, 839—846 (1951).

Reznik, H.: Beiträge zur Physiologie der Verholzung. Planta **54**, 333—364 (1960).

Riley, R., Bennet, M. D.: Meiotic DNA synthesis. Nature **230**, 182—185 (1971).

Roberts, R. M., Butt, V. S.: Patterns of cellulose synthesis in maize root-tips. A chemical and autoradiographic study. Exp. Cell Res. **46**, 495—510 (1967).

Roberts, R. M., Butt, V. S.: Patterns of incorporation of pentose and uronic acid into cell walls of maize root-tips. Exp. Cell Res. **51**, 519—530 (1968).

Roberts, R. M., Butt, V. S.: Patterns of incorporation of D-galactose into cell wall polysaccharide of growing maize roots. Planta **84**, 250—262 (1969).

Rogers, A. W.: The photometric measurement of grain densities in autoradiographs. J. Physiol. **190**, 14 P (1967).

Rogers, A. W.: Techniques of autoradiography, 338 p. Amsterdam, London, New York: Elsevier Publ. Comp., 1969.

Ron, A., Prescott, D. M.: The efficiency of tritium counting with seven radioautographic emulsions. In: Methods in cell physiology (D. M. Prescott, ed.), Vol. IV, 231—240. New York, London: Academic Press 1970.

Rosenberger, R. F., Kessel, M.: Nonrandom sister chromatid segregation and nuclear migration in hyphae of *Aspergillus nidulans*. J. Bact. **96**, 1208—1213 (1968).

Ross, C., Cole, C. V.: Metabolism of cytidine and uridine in bean leaves. Plant Physiol. **43**, 1227—1231 (1968).

Ruddel, C. L.: Hydroxyethyl metacrylate combined with polyethylene glycol 400 and water; an embedding medium for routine 1—2 micron sectioning. Stain Technol. **42**, 119—123 (1967).

Ruthmann, A.: Methods in cell research. New York: Cornell Univ. Press Ithaca 1970.

Sadava, D., Chrispeels, M.: Cell wall protein in plants. Autoradiographic evidence. Science **165**, 299—300 (1969).

Sagan, L.: An unusual pattern of tritiated thymidine incorporation in *Euglena*. J. Protozool. **12**, 105—109 (1965).

Sagan, L., Ben-Shaul, Y., Schiff, J. A., Epstein, H. T.: Radiographic localization of DNA in the chloroplasts of *Euglena*. J. Cell Biol. **23**, 81 A (1965).

Sauter, J. J.: Autoradiographische Untersuchungen zur RNS- und Proteinsynthese in den Sporenmutterzellen, jungen Pollen und Tapetumzellen während der Mikrosporogenese von *Paeonia tenuifolia* L. Z. Pflanzenphysiol. **61**, 1—19 (1969).

Savage, J. R. K., Wigglesworth, D. J.: Labelling microspores with tritiated thymidine by bud injection in *Tradescantia*. Stain Technol. **46**, 37—39 (1971).

Sawicki, W., Darzyniciewicz, Z.: Polyvinyl chloride protective films in autoradiography. Folia Histochem. Cytochem. **1**, 283—288 (1963).

Sawicki, W., Pawinska, M.: Effect of drying on unexposed autoradiographic emulsion in relation to background. Stain Technol. **40**, 67—68 (1965).

Sawicki, W., Ostrowski, K., Rowinski, J.: Effect of exposure temperature on overall efficiency of autoradiographs and on fogging of emulsion. Stain Technol. **43**, 35—39 (1968).

Schneider, L. K., Rieke, W. O.: DNA replication patterns and chromosomal protein synthesis in opossum lymphocytes in vitro. J. Cell Biol. **33**, 497—509 (1967).

Scott, S. N., Shah, V. C., Smillie, R. M.: Synthesis of chloroplast DNA in isolated chloroplasts. J. Cell Biol. **38**, 151—157 (1968).

Setterfield, G., Bailey, S. T.: Deposition of cell walls in oat coleoptiles. Canad. J. Bot. **37**, 861—870 (1959).

Setterfield, G., Schreiber, R., Woodard, J.: Mitotic frequency determination and microphotometric feulgen dye measurement in root tips. Stain Technol. **29**, 113—120 (1954).

Shapiro, I. M., Polikarpova, S. I.: Autoradiographic study on the synthesis of nuclear proteins. VI. The data on the turnover rate of nuclear proteins and formation of the protein component in chromosomes of chinese hamster cells cultured in vitro. Chromosoma **28**, 188—198 (1969).

Shephard, D. C.: An autoradiographic comparison of the effects of enucleation and actinomycin D on the incorporation of nucleic acid and protein precursors by *Acetabularia* chloroplasts. Biochem. Biophys. Acta **108**, 635—643 (1965).

Sisken, J. E.: The synthesis of nucleic acids and proteins in the nuclei of *Tradescantia* root tips. Exp. Cell Res. **16**, 602—614 (1959).

Sprey, B.: Zum Verhalten DNS-haltiger Areale des Plastidenstromas bei der Plastidenteilung. Planta **78**, 115—133 (1968).

Steffensen, D. M.: Proteins of sperm nuclei examined by autoradiography at fertilization and subsequent nuclear division in *Lilium*. Genetics **52**, 631—641 (1965).

Steffensen, D. M.: Synthesis of ribosomal RNA during growth and division in *Lilium*. Exp. Cell Res. **44**, 1—12 (1966).

Steffensen, D. M., Sheridan, W. F.: Incorporation of 3H-Thymidin into chloroplasts DNA of marine algae. J. Cell Biol. **25**, 619—626 (1965).

Stein, O., Stein, Y., Goodman, D. S., Fidge, N. H.: The metabolism of chylomicron cholesteryl ester in rat liver. J. Cell Biol. **43**, 410—431 (1969).

Stevens, G. W. W.: Radioactive microphotographs for resolution testing in autoradiography. Brit. J. Radiol. **23**, 723 (1950).

Stocking, C. R., Gifford, E. M., Jr.: Incorporation of thymidine into chloroplasts of *Spirogyra*. Biochem. Biophys. Res. Commun. **1**, 159—164 (1959).

Stöcker, E., Müller, H. A.: Zur chemischen Induktion von Silberkörnern im Strippingfilm bei der Orcein-Quetsch-Technik. Histochemie **11**, 167—170 (1967).

Stubblefield, E.: Quantitative tritium autoradiography of mammalian chromosomes. I. The Basic Method. J. Cell Biol. **25**, 137—147 (1965).

Suyama, Y., Eyer, J.: Ribonucleic acid synthesis in isolated mitochondria. J. Biol. Chem. **243**, 320—328 (1968).

Tamm, C., Hodes, M. E., Chargaff, E.: The formation of apurinic acid from the desoxyribonucleic acid of calf thymus. J. Biol. Chem. **195**, 49—63 (1952).

Taylor, J. H.: Autoradiographic detection of incorporation of P^{32} into chromosomes during meiosis and mitosis. Exp. Cell Res. **4**, 169—179 (1953).

Taylor, J. H.: The mode of chromosome duplication in *Crepis capillaris*. Exp. Cell Res. **15**, 350—357 (1958a).

Taylor, J. H.: Sister chromatid exchanges in tritium labelled chromosomes. Genetics **43**, 515—529 (1958b).

Taylor, J. H.: Incorporation of phosphorus-32 into nucleic acids and proteins during microgametogenesis of *Tulbaghia*. Amer. J. Bot. **45**, 123—130 (1958c).

Taylor, J. H.: Autoradiographic studies of nucleic acids and proteins during meiosis in *Lilium longiflorum*. Amer. J. Bot. **46**, 477—484 (1959).

Taylor, J. H.: Distribution of tritium-labelled DNA among chromosomes during meiosis. I. Spermatogenesis in the grasshopper. J. Cell Biol. **25**, 57—67 (1965).

Taylor, J. H., Woods, P. S., Hughes, W. L.: The organization of chromosomes as revealed by autoradiographic studies using tritium labelled thymidine. Proc. Nat. Acad. Sci. (Wash.) **43**, 122—128 (1957).

Thurston, J. M., Joftes, D. L.: Stains compatible with dipping radioautographs. Stain Technol. **38**, 231—235 (1963).

Tsai, C. S., Axelrod, B.: Catabolism of pyrimidines in rape seedlings. Plant Physiol. **40**, 19—44 (1965).

Vanha-Perttula, T., Grimley, P. M.: Loss of protein and other macromolecules during preparation of cell cultures for high resolution autoradiography. J. Histochem. Cytochem. **18**, 565—573 (1970).

Villadsen, I. St., Zeuthen, E.: Synchronization of DNA synthesis in *Tetrahymena* populations by temporary limitation of access to thymine compounds. Exp. Cell Res. **61**, 302—310 (1970).

Waller, C.: The properties of nuclear emulsions in relation to autoradiography. 1st european symposium autoradiography, Rome 1961. (quoted from Rogers, 1969).

Wand, M., Zeuthen, E., Evans, E. A.: Tritiated thymidine: Effect of decomposition by self-radiolysis on specifity as a tracer for DNA synthesis. Science **157**, 436—438 (1967).

Wardrop, A. B.: Cell-wall formation in root-hairs. Nature **184**, 996—997 (1959).

Wardrop A. B., Harada, H.: The formation and structure of cell wall in fibres and tracheids J. exp. Bot. **16**, 356—364 (1965).

Wimber, D. E.: Duration of the nuclear cycle in *Tradescantia paludosa* root tips as measured with H^3-thymidine. Amer. J. Bot. **47**, 828—834 (1960).

Wimber, D. E.: Asynchronous replication of deoxyribonucleic acid in root tip chromosomes of *Tradescantia paludosa*. Exp. Cell Res. **23**, 402—407 (1961).

Wimber, D. E.: Duration of the nuclear cycle in *Tradescantia* root tips at three temperatures as measured with H^3-thymidine. Amer. J. Bot. **53**, 21—24 (1966).

WIMBER, D. E., PRENSKY, W.: Autoradiography with meiotic chromosomes of the male newt (*Triturus viridescens*) using 3H-thymidine. Genetics **48**, 1731—1738 (1963).

WOLFF, S.: Are sister chromatid exchanges sisterstrand crossovers or radiation induced exchanges? Mutation Res. **1**, 337—343 (1964).

WOLFF, S.: Handling of root tips. In: Methods in Cell Physiology, ed. by D. M. PRESCOTT, Vol. I, p. 215—217. New York-London: Academic Press 1964.

WOLFF, S., LUIPPOLD, H. E.: Chromosome splitting as revealed by combined X-ray and labelling experiments. Exp. Cell Res. **34**, 548—556 (1964).

WOLLGIEHN, R., MOTHES, K.: Über die Inkorporation von 3H-Thymidin in die Chloroplasten-DNS von *Nicotiana rustica*. Exp. Cell Res. **35**, 52—57 (1964).

WOODARD, J. E., RASCH, E., SWIFT, H.: Nucleic acids and protein metabolism during the mitotic cycle in *Vicia faba*. J. Biophys. Biochem. Cytol. **9**, 445—462 (1961).

WOODCOCK, C. L. F., BOGORAD, L.: Evidence for variation in the quantity of DNA among plastids of *Acetabularia*. J. Cell Biol. **44**, 361—375 (1970).

WOODS, P. S., TAYLOR, J. H.: Studies of ribonucleic acid metabolism with tritium-labelled cytidine. Lab. Invest. **8**, 309—318 (1959).

ZAJICEK, G., GROSS, J.: Studies on the radiobiological effect of ^{3}H-thymidine on the DNA turnover in Landschütz ascites tumour cells. Exp. Cell Res. **34**, 138—143 (1964).

ZILKEN, J.: Zeitpunkt der Chromosomenreduplikation während der Meiose bei *Sphaerocarpos*. Staatsexamenarbeit, Heidelberg, 17 S. (1971).

Chapter 6

Autoradiography with the Electron Microscope: Experimental Techniques and Considerations Using Plant Tissues

J. D. Pickett-Heaps
Department of Molecular, Cellular and Development Biology
University of Colorado
Boulder, Colorado, 80302, U.S.A.

Contents

6.1. *Introduction*

Considerable interest in autoradiography has been stimulated by the development of this technique for electron microscopy. However, the subsequent upsurge in enthusiasm has often resulted in overconfidence in the technique and sometimes an unfortunate lack of discrimination in the interpretation of its results. This has led to a profound and often justified scepticism among many more discerning workers who recognize the severe limitations of the methods used today. Autoradiography is not a panacea which will make possible the identification of any stray cytoplasmic vesicle or inclusion; like any other scientific method it can only yield valuable results when its limitations are clearly recognized, stated and evaluated where possible. More recently, several elegant and important autoradiographic experiments have been published illustrating these limitations and methods of evaluating them (see Section 6.15).

This chapter will attempt three things. Firstly, it will introduce any worker unfamiliar with electron microscopic autoradiography to various relatively simple experimental procedures which should enable good autographic results to be obtained. If these results then come up to expectation and warrant further development, the worker concerned will probably wish to try out various more sophisticated methods, so secondly, there will be a brief summary of various recent advances and elaborations in techniques. Thirdly some drawbacks in the methods of autoradiography, and particularly of the problems of interpretation will be briefly mentioned. In my opinion, the techniques although tedious and painstaking, are relatively straightforward and good electron-microscopic autoradiograms are *not* difficult to obtain; what appears vastly more difficult is to know what these autoradiograms really mean, and to design experiments which will give clear-cut unambiguous information.

6.2. *Choice of Tissue*

This of course will be often determined by the nature of the investigation. However, it is of considerable advantage to use where possible, small pieces or regions of tissue; this ensures that the radioactive metabolite administered is not diffused throughout too large a number of cells, and so increases the amount of incorporation per cell and consequently per thin section. A root tip (used intact or excised) is perhaps the easiest whole tissue to use initially if possible, since the actively growing region can be immersed in a small drop of relatively concentrated metabolite placed on a hydrophobic surface (e.g. polythene).

Furthermore, clearest results are likely to be obtained using specialized cells or tissues actively engaged in specific functions. For example, polysaccharide secretion can easiest be studied in those cells whose metabolism is largely directed towards the laying down of massive cell walls, or in cells (e.g. of the root cap) whose function involves secretion of mucilage (NORTHCOTE and PICKETT-HEAPS, 1966); in comparison, undifferentiated cells (e.g. in the root meristem) are engaged in many activities and equivalent studies on such cells may be correspondingly ambiguous and less easy to interpret (e.g. wall deposition — PICKETT-HEAPS, 1967). Successful utilisation of autoradiography is often consequent upon careful choice of experimental tissue.

6.3. *Choice of Metabolite*

The choice of metabolite will be determined mainly by the availability of suitable labeled compounds and by biochemical considerations. Quite obviously, if the fate of a labeled metabolite has already been followed biochemically, this is of considerable value in the interpretation of experimental autographic results; indeed most research projects utilising autoradiography should attempt some biochemical investigations into the fate of administered compounds. However, even these control experiments can be misleading. Only those labeled derivatives that are bound by fixation (Section 6.7.1) and that are quite insoluble during fixation, dehydration and embedding are likely to be retained in the thin section finally used for autoradiography (see BERGERON and DROZ, 1968; DERMER, 1968). This means that detailed biochemical data regarding the labeling of soluble intermediates in biochemical sequences is probably somewhat irrelevant. Furthermore, some metabolites that biochemical consideration would indicate might be useless, may in fact prove valuable. For example, in root tips tritiated glucose could be expected to be broken down very rapidly, with the label being disseminated into a wide range of derivatives; however, it has been useful in studies of cell wall formation (e.g. Northcote and Pickett-Heaps, 1966; Pickett-Heaps, 1966, 1967; Wooding, 1968 et al.), probably because i) the cells examined were engaged to a large extent in polysaccharide synthesis (see Section 6.2); ii) soluble intermediates were most conveniently lost during processing; and iii) the administered glucose very rapidly became incorporated into insoluble polysaccharides whereas the incorporation of labeled breakdown products into other cell constituents (e.g. proteins, nucleic acids) was much slower. Thus, there is no easy way to judge whether any particular metabolite will yield a useful result. My opinion is that biochemical analyses of labeled tissues must be performed if possible, but they should follow the establishment of a successful autoradiographic experiment when the effort involved is known to be worthwhile. Ideally, these biochemical analyses should be performed on tissues that have at least been fixed and dehydrated if this does not complicate the biochemical results too much. Oxidative fixations etc. may degrade certain polysaccharides or other cytoplasmic derivatives more easily than others, so that the labeling detected in the autographic image may represent a more restricted class of compounds than expected. For example, relevant again to studies of cell wall structure and synthesis, it is known that pectic materials may be rapidly degraded under neutral or mildly alkaline conditions (Albersheim, Neukom and Deuel, 1960).

As regards the availability of labeled metabolites, certain firms (e.g. The Radiochemical Centre, Amersham, Bucks, England) will make special compounds to order if feasible. The extra cost involved may not be excessive and could be justified if a really specific precursor is needed. For example, I have had cinnamic acid specially tritiated to a high specific activity for use in studies on lignification (this compound is incorporated into lignin readily and into very little else). In autoradiograms, it was taken up in very large amounts specifically into xylem wall thickenings as hoped (Fig. 6.8 — Pickett-Heaps, 1968).

6.4. *Choice of Radioactive Isotope*

A variety of radioactive isotopes have been successfully used in electron microscopic autoradiography (e.g. P^{32}: Caro and Schnos, 1965; I^{125}: Maunsbach, 1966; S^{35}: Fewer, Threadgold and Sheldon, 1964; Fe^{55}: Orlic, 1968). For the labeling of organic precursors, however, tritium remains the obvious choice of electron-microscopists. Its half life (*ca.* 31 years) is not short enough to complicate experiments. Of considerable importance is the availability of tritiated metabolites of very high specific activity. This means that cells can be very highly labeled since a high proportion of incorporated molecules will contain tritium, whilst the very low concentration of the metabolite will probably not affect the tissues' general metabolism (see below).

Autoradiography is a very insensitive method of detecting radioactivity partly because ultrathin sections contain so little of the original tissue, and it is of paramount importance to get as much radioactivity as possible into these sections; otherwise the photographic exposure period of the emulsion may last many months, rendering further development of the experiment tedious or even causing abandonment of the whole affair. There is a possibility of radiation damage becoming appreciable in heavily labeled tissues (e.g. incorporation of large amounts of tritium into DNA), but with fairly short exposures of live tissues to most metabolites, this problem is usually of lesser importance.

Another most important reason for choosing tritium is related to the low energy of emission upon radioactive decay (0.013 mev) of the electron which exposes the nearby silver halide crystal in the autogram. This low energy renders the electron's capture in the halide crystal easy and, more important, ensures that crystals cannot be exposed very far from the point of decay; this latter consideration affects the resolution achievable (CARO, 1962).

Various aspects concerning resolution are discussed in many more theoretical papers and books (e.g. CARO, 1962); CARO and SCHNOS (1965) also compare the resolution achievable using H^3 and P^{32} (they quote 0.1 and 0.3 microns respectively under their experimental conditions).

6.5. *Administration of the Metabolite*

The method used to administer the labeled metabolite will be dictated largely by the tissue used. Tissues should be damaged as little as possible for obvious reasons; root tips again are excellent tissues to use since they take up many metabolites directly through their cell walls and they can almost always be used intact. Many tissues will however have to be excised and placed into the solution. Uptake of metabolites into excised wheat coleoptiles (PICKETT-HEAPS, 1966, 1968) and stems of pine and sycamore seedlings (WOODING, 1968) has been found adequate although many cells must have been killed at the excision region. Such damage creates problems in the study of callose deposition in phloem tissue for example (see NORTHCOTE and WOODING, 1966). It may be preferable not to use segments of tissue;

whole excised coleoptiles or short stems can be floated on the solution of metabolite; for fixation, they should be placed in the fixative, a very thin slice of tissue at the excision region should then be discarded and the next slice (*ca* 1 mm) can then be retained in the fixative for radioautography. (For all excisions, make sure the razor or scalpel blade is grease-free). The problem of bacterial contamination of the labeled metabolite may become important. It is difficult to prepare sterile seedlings etc. and autoradiograms will generally soon show that bacteria on the tissue surfaces readily incorporate the administered metabolite. Provided bacterial contamination is not excessive and the incubations are not too long, it is probably best just to rinse the tissue in sterile water a couple of times prior to incubation, and accept some such contamination. Inclusion of antibiotics with the labeled metabolite may have pronounced and subtle effects on the tissues' metabolism which are difficult to evaluate, so these measures merit careful consideration.

The concentration of labeled metabolite needed for good results can only be estimated from trial experiments. I routinely make up tritiated compounds to a final specific activity around 0.5—2.0 mC/ml and up to 5 mC/ml for short pulse experiments. This is very high by normal standards, but is usually necessary in order to avoid prolonged photographic exposure of the emulsion (see comments in Section 6.4).

6.6. *Pulse Labeling*

The biochemical technique of pulse labeling sometimes may be used to great advantage in autoradiography, since a movement of a labeled derivative may be inferred from the sequential appearance of radioactivity in different parts of the cell. Autoradiography of root cap cells using pulse labeling techniques with tritiated glucose showed an initial accumulation of labeled derivatives within the golgi bodies (Fig. 6.5) which was followed during the cold chase by the disappearance of label from these organelles and instead, the accumulation of radioactivity in the slime layer (Northcote and Pickett-Heaps, 1966). Whilst another interpretation is remotely possible (Section 6.16.3), this experiment demonstrated unambiguously the direction of movement of polysaccharide through the cell.

Pulse labeling may require some thought. Firstly, it is often not obvious how long the exposure of the tissue to the metabolite need be to ensure sufficient labeling of the autograms (with root cap cells, 5—10 min sufficed — Fig. 6.5). This period can only be determined by trial and error. Secondly, if the concentration of the cold metabolite is high enough to ensure adequate dilution of the remaining labeled metabolite within the tissue, sometimes it may also affect the tissue appreciably. For example, whilst a very low concentration of labeled cinnamic acid did not discernably affect wheat xylem cells, the subsequent chase using a higher concentration of cold cinnamic acid disrupted somewhat the formation of the xylem wall thickenings being studied (PICKETT-HEAPS, 1968). In another example, it appeared that the balance between metabolic pathways in root cap cells was altered by the cold glucose chase that followed a pulse of tritiated glucose (NORTHCOTE and PICKETT-HEAPS, 1966). The actual concentration of labeled glucose used was low and derivatives were concentrated in the golgi bodies but not in the starch grains; since starch was labeled in other cells (including the inner cap cells) it was inferred that in outer cap cells, starch was probably being broken down to provide glucose residues for polysaccharide synthesis in the golgi bodies. However, when a higher concentration (1 mgm/ml) of cold glucose was used as a chase after a fairly long ($^{3}/_{4}$—$1^{1}/_{4}$ hrs) incubation in tritiated glucose, the golgi bodies of cap cells soon lost most of their transient labeling as expected, but now the starch grains had become heavily labeled. So it appeared that during the cold chase, sufficient glucose was available for the cells to undertake both polysaccharide *and* starch synthesis, the latter then slowly accummulating radioactivity from the various pools of glucose and derivatives (whose radioactivity would be becoming diluted during the chase). This cold chase therefore seemed to have disturbed the balance between various biochemical pathways functioning within the cells.

6.7. *Fixation*

After administration of the radioactive label, the tissue is fixed by usual methods but some care is needed in the choice of fixative.

6.7.1. — Glutaraldehyde Fixation

Glutaraldehyde (followed with osmium) has been generally established as the most valuable fixative to date for most types of cells and tissues. Its' use is recommended in autoradiographic experiments with one very important qualification. Glutaraldehyde has two reactive aldehyde groups per molecule, which by cross linking in and between biological molecules fixes cytoplasm in a stable form; however, whilst one of these aldehyde groups is bound in the cytoplasm, the other in some proportion of the glutaraldehyde molecules could theoretically react with free reactive groups in the radioactive metabolite (if present), which would lead to quite unspecific binding of the administered metabolite to the cytoplasm. That this happens with amino-acids has been demonstrated by the important results of Peters and Ashley (1967) and Bergeron and Droz (1968). Peters and Ashley found appreciable autographic labeling of glutaraldehyde- or osmium-fixed liver cells exposed to tritiated leucine in the presence of puromycin, a drug which allows the absorption of the leucine but prevents its incorporation into proteins. These results agree with those of Bergeron and Droz (1968) who followed also the extraction of radioactivity during fixation and further processing. Formaldehyde has been indicated as the fixative of choice following administration of amino-acids, and the importance of controls in these cases was stressed.

For fixation itself, I have seldom had reason to believe that the detailed composition of fixatives etc. is critical. Glutaraldehyde works well at concentrations between $^1/_2$—6% (and probably outside these limits). Its pH should be around neutrality; I use it between pH 6.4—7.8, buffering with phosphate or other buffers (incidentally, glutaraldehyde is reported to react quantitatively with veronal acetate); it can work well unbuffered too, if the pH of its solution is brought to neutrality. My main concern is that the fixative's osmolarity should not be high enough to plasmolize the cells. I also include a low concentration of calcium which might possibly assist in membrane and microtubule preservation. Fixation normally lasts between $^1/_4$—2 hrs, but can be much longer (e.g. 48 hrs or more). In summation, I recommend about 2% glutaraldehyde, buffered to pH 7 with 0.025 M sodium/potassium phosphate buffer containing about 0.002 M $CaCl_2$ (ignore any faint precipitate). Specimens should *not* be prechilled and then fixed at 0—6 °C without good reason, since microtubules break down in cells maintained at low temperatures.

Before post-fixation, excess glutaraldehyde should be removed from tissues as it slowly reacts with osmium. Several rinses in the buffer and then one or two soakings of the tissue blocks for $^1/_2$—1 hr will suffice.

Post-fixation with osmium is straightforward. A suitable concentration is 1% buffered to neutrality with veronal-acetate or phosphate (made up as for glutaraldehyde, containing calcium). Normal times of post-osmification vary from 1—2 hrs at room temperature to overnight at 0 °C (once stabilised by fixation, microtubules of course no longer dissociate at low temperatures). Finally, excess osmium is washed out with several rinses of water (it will otherwise precipitate out during dehydration).

As I have indicated, compact meristematic plant cells are easy to fix well (hence their popularity in electronmicroscopical studies). Kidwai and Robards (1969) believe that vascular tissue in general needs carful treatment, but in my experience wheat xylem cells are easy to preserve well.

Sieve-tubes always pose special problems which cannot be described here (e.g. see Cronshaw and Anderson, 1969). Many algal cells are relatively easy to fix but, until recent improvements in techniques, have proved impossible to embed (see Section 6.8).

6.7.2. — Permanganate Fixation

Although permanganate fixation is seldom used nowadays because it is very destructive, occasionally it may be valuable. The main reasons it has been successful in investigations into cell wall deposition in plants (e.g. Northcote and Pickett-Heaps, 1966), seem to be as follows: i) it gives very clear images of certain membraneous organelles (e.g. the golgi bodies); ii) although polysaccharides may be partly oxidised by permanganate, this degradation cannot be excessive since much labeled material proved to be present in sections (e.g. Fig. 6.5); iii) the severely destructive effect of permanganate towards other cytoplasmic components may be valuable (i.e. if one is not interested in them). For example, if both proteins and polysaccharides were to be labeled, permanganate fixation would probably destroy most labeled proteins, so yielding a much clearer picture of an incorporation into polysaccharide.

As for experimental details, I usually use 1—4% permanganate, unbuffered, for between $^1/_2$—2 hrs at room temperature; I also include 0.002 M $CaCl_2$ but cannot vouch for its necessity. The tissue blocks are well washed before dehydration.

6.8. *Dehydration and Embedding*

After fixation, dehydration and embedding follow as normal, during which some loss of radioactivity might occur, particularly from labeled lipids (Dermer, 1968; Cope and Williams, 1968; Stein et al., 1969). As with fixation, compact meristematic plant cells pose few problems at this stage. They can be quite rapidly run through a graded series of ethanol or acetone in water. Final dehydration however, should be thorough because many resins are susceptible to traces of water, and will refuse to harden properly if any remains; several changes of absolute ethanol over approximately 12—24 hrs are therefore recommended. Before embedding in the resin selected, I prefer to replace ethanol by propylene oxide (several changes) which is more volatile and so evaporates away completely during curing of the resin. Many recipes for embedding media exist. I have always found Araldite suitable for plant tissues and Spurr's (1969) new low-viscosity medium can be highly recommended.

When one is interested in very vacuolated or indeed any non-meristematic higher plant cells, various problems may be encountered during embedding (e.g. collapsed cells, refractory cell walls etc.). Workers therefore should be familiar with more unusual tissues and their problems before attempting autoradiography using standard preparative methods. Such problems are not insoluble: for example, delicate algal cells such as *Oedogonium* and *Spirogyra* have now been successfully embedded in resin using modified (and vastly more tedious) techniques (e.g. Pickett-Heaps and Fowke, (1969) and recently we have now even managed to embed well-preserved giant coenocytic cells (Marchant and Pickett-Heaps, 1970).

6.9. *Preparation of Sections for Autoradiography*

Sections of labeled tissues should be cut slightly thicker than normal for autoradiography (i.e. up to gold interference colours). Resolution of very fine detail in the micrographs is seldom necessary and thicker sections contain more radioactivity and give better contrast, often needed because of the various layers of carbon, Parlodion, gelatine etc. on each side of the sections. However, thicker sections will also decrease resolution (Caro, 1962). In any event, inclusion of sections of varying thicknesses on each grid is easy and enables the best results to be chosen. Sections can either be mounted on bare grids, or on larger mesh grids covered with a convenient support film. I have not encountered undue contamination from photographic processing of specimens mounted thus on copper grids, provided care was exercised. (Salpeter and Bachmann, 1964, prefer to do all their film coating and later photographic processing before mounting the sections on grids). Staining is best done next; some workers have successfully stained after final photographic processing, but this is not recommended at least in initial attempts of autoradiography. Lead or uranium stains do not apparently affect the sensitivity of the emulsion (e.g. by quenching); neither does uranium significantly add to the background exposure.

Before coating the sections with emulsion, they must first be coated with carbon (unless the worker prefers to stain after photographic processing). The carbon coat has two functions. Firstly, it prevents the photographic developing and fixing solutions reacting with the heavy metal stains in the section (which results in either contamination or more generally, an efficient removal of all the stains and consequently the image of

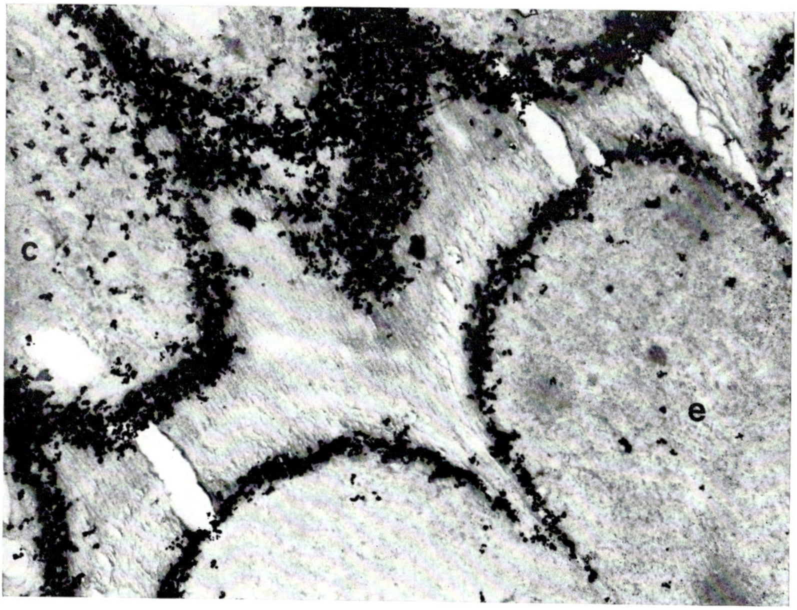

Fig. 6.1. Transverse section, root tip exposed to glucose-6-H^3 for ½ hour; permanganate fixed. The section was not coated with carbon (see Section 6.9) and so subsequent photographic processing removed heavy metals from the section, leaving a "ghost image" — note tears in the section. Silver grains are concentrated at the outer edge of epidermal cells (e) and over the secretion of the nearby root cap cells (c). × 3,500

the cells; Fig. 6.1; Koehler, Mühlethaler and Frey-Wyssling, 1963). Secondly, some emulsions react with osmium in the sections, causing a loss of the latent image during exposure (Salpeter and Bachmann, 1964). A layer of carbon interposed between sections and emulsion will prevent both these effects.

6.10. *Choice of Emulsion*

There are many emulsions now available for electronmicroscopic autoradiography. The two perhaps most commonly used are Ilford L4 and Kodak NTE. Ilford L4 is very easy to handle and is recommended for first attempts at autoradiography. Kodak NTE has a number of drawbacks (e.g. too much gelatine in the emulsion, latent image fading in the film etc. — Salpeter and Bachmann, 1964 et al.) and its' use should be reserved until some experience is gained in the methods.

6.11. *Coating the Sections*

There are many methods to choose from in the literature. I will simply outline two, one fairly crude, which are both quite easy. The aim is to coat the sections with a film consisting of a monolayer of close packed silver halide crystals (see Caro and van Tubergen, 1962). Earlier coating techniques often gave a considerable variation in film thickness, but modern methods yield films which are generally very good.

6.11.1. — Loop Method

This rather crude method, developed by Caro and van Tubergen (1962) and others, often yields good results and is very simple to carry out. Ilford L4 emulsion is melted under safe lights at about 40 °C in a water bath with about 2—3 times its' weight in distilled water. After gentle stirring for about 15 min, it is carefully decanted into another beaker and cooled. Caro and van Tubergen recommend holding it at 0 °C for a couple of minutes, then allowing it to reach room temperature for $^1/_2$ hr. (My experience has been that this preliminary cooling in an ice bath has little effect on the properties of the film formed.) Then a loop (3—5 cms. in dia.) is immersed in the emulsion and slowly withdrawn vertically, forming a film. Caro and van Tubergen say this film gels almost immediately. I have seldom found this myself and recently Caro (1969) suggests a method to promote such gelling; however, the method works reasonably well even if gelling does not occur. The film is then touched on to the surface of the grid and broken.

An examination of some such coated grids in the electronmicroscope (without photographic development) as a control will reveal the distribution of silver halide crystals; such a check should be made on all batches of grids coated with emulsion, regardless of the method used. Using this crude loop method and adjusting the dilution of the emulsion to get a thin enough layer, the films formed are often good, certainly good enough for many preliminary experiments. However, some uneveness in distribution must be expected; for example the halide crystals will often be much thicker near grid bars than at the centre of the grid holes (Fig. 6.2). Furthermore, since the film is put on to the grid whilst still quite labile, halide crystals can move after coating, often away from sections themselves, leaving some areas sparsely coated (Fig. 6.3).

Fig. 6.3. Silver grains (Ilford L-4 emulsion) placed over a section by the loop method (Section 6.11.1); the grains have covered the section irregularly and in a nonrandom fashion. × 7,000

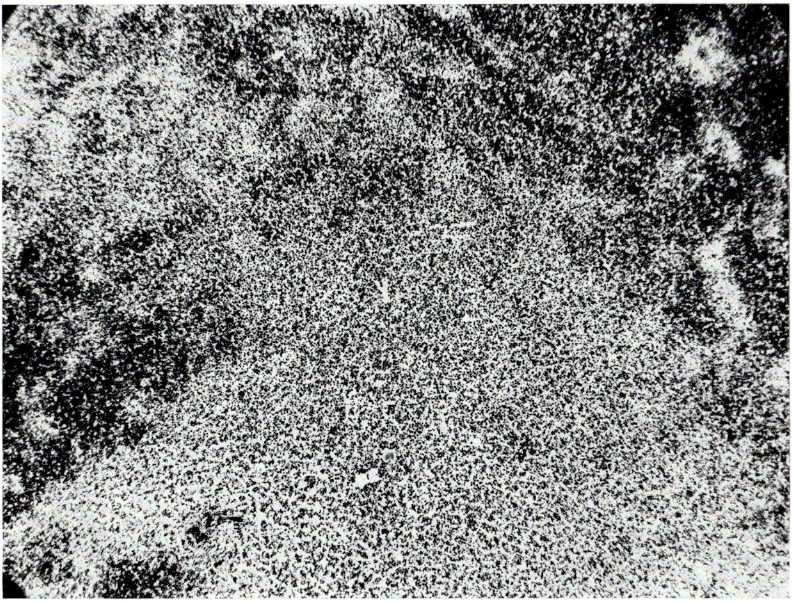

Fig. 6.2. Survey of the distribution of silver-halide crystals placed on a coated grid by the loop method (Section 6.11.1). The crystals form quite a good monolayer at the centre of the grid hole, but they are uneven and thick at the edges. × 1,000 approximately

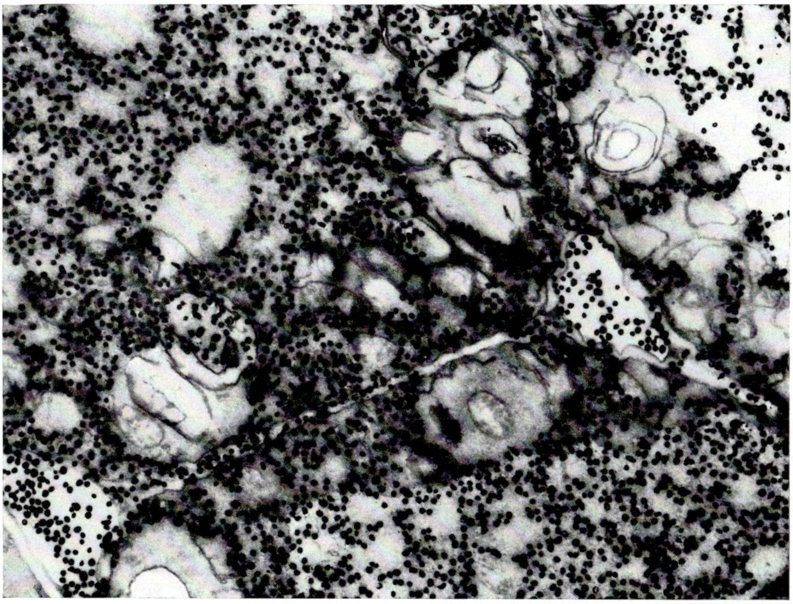

6.11.2. — The "Agar-Block" Method

To overcome these problems and improve reproducibility, CARO and VAN TUBERGEN (1962) developed their ingenious "agar-block" technique. I can strongly recommend this excellent method, particularly with L4 emulsion. Although slightly difficult and tedious on a first attempt, a little practice renders it very straightforward and reproducible. The method given below is only slightly modified from the original.

A strip (e.g. $11'' \times 1^3/_4'' \times {}^1/_4''$) of glass with perfectly clean surfaces is placed in a long, flat trough of stainless steel (e.g. $12'' \times 2^1/_2'' \times {}^3/_4''$) and both are warmed to about 60 °C. A 2% solution of highest grade agar (e.g. Difco Agar Noble) is autoclaved, cooled to 60 °C and poured over the glass to fill the trough, where it is allowed to set. When at room temperature, the agar around the edges of the glass is carefully cut away and removed; then the glass strip (covered with the slab of agar) is also removed from the trough. The agar is then sliced into blocks (e.g. up to $1'' \times 1^1/_2''$). To remove these blocks from the glass strip, a clean microscope slide is carefully lowered on to the top of each one till block and slide adhere quite firmly; gentle prising now with a razor blade at one edge of the block/strip interface accompanied by gentle twisting of the slide breaks the seal of moisture. The agar blocks can then be removed from the strip, whilst being held on the microscope slide; their edges can be retrimmed with a razor.

The film of moisture on the flawless (and dust-free!) surface of these agar blocks is removed by drying (e.g. brief exposure to an infra-red lamp). Then each block surface is flooded with a 0.2% solution of Parlodion in amyl acetate and dried vertically, after which the blocks are removed to the darkroom. The Parlodion films will later be stripped off the agar by floating them on to a water surface; at this stage it is a wise precaution to strip off at least one of the series in the light to check that all is well with the films.

All necessary items are now transferred to the darkroom for preparation of the autographic film. A complete trial run-through of the method in the light beforehand is recommended and will reveal what minor equipment is needed to carry out the next stages (e.g. beakers, thermometers, forceps etc.). This will prevent the otherwise inevitable and infuriating lack of some vital but minor utensil when the real thing is under way with emulsions etc. under safe lights.

In the darkroom under appropriate safe lights, Ilford L4 emulsion, diluted about 1 : 4 with water, is melted at 45 °C. After some minutes it is decanted and cooled to room temperature; as before, a film is formed from the diluted emulsion with a loop large enough to fit over the agar blocks. This film is carefully lowered on to the parlodion film on the agar and then broken (i.e. around the edge of the blocks). Since the surface of the parlodion is perfectly flat and of a large area (see also GRANBOULAN, 1965: p. 49), halide grains will not move and close-packed mono-layers of them can be easily formed. Furthermore, the distribution of silver grains is completely unaffected by the later presence of sections underneath (Fig. 6.4 — compare with Fig. 6.3). If necessary, the original dilution of the L4 emulsion can be adjusted.

If now the agar blocks are lowered into a water bath at an angle, the parlodion film plus the monolyer of silver crystals can be stripped off; this rarely causes trouble and the film is quite visible by red light on the water surface. Grids are coated with this film as normal. They can either be placed face down on the floating film and the whole lot removed (e.g. with filter paper) or else they can be placed underwater on a wire mesh which is raised up under the film.

Fig. 6.4. Film of Ilford L-4 emulsion (purposely rather thin in this case) prepared by the Agar-block method (see Section 6.11.2) and placed over a very thick, irregular section. The distribution of silver grains is even and random with respect to the section (c.f. Fig. 6.3). × 3,200

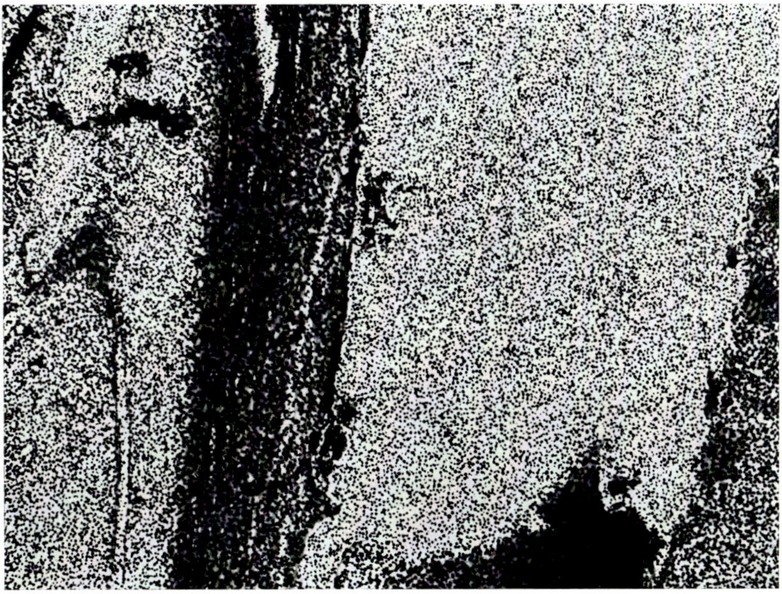

6.12. *Storage*

The coated grids should be kept in a cool, dry environment (e.g. small light-tight boxes containing silicagel, appropriately stored). L4 emulsion seems unaffected by atmospheric oxygen, and so storage in an inert atmosphere (needed with NTE emulsions — Salpeter and Bachmann, 1964) is unnecessary. The required exposure period can be found by processing and examining some of the grids at intervals.

6.13. *Development and Fixation*

Grids can most conveniently be processed by placing them on a large drop of each reagent, freshly made up. I routinely develop the image in D-76, D-19 or Microdol for 1—2 min (Caro and van Tubergen, 1962, recommend up to 4 min), after which the grids are placed on 1% acetic acid for about 10 sec, and then on a rapid fixer for 1—2 min. They are carefully and thoroughly rinsed and soaked 5 min before being dried and examined. Underdevelopment is not recommended; the resultant smaller grains may only create the illusion of greater resolution (see Granboulan, 1965: p. 51).

6.14. *Removal of Gelatine from Sections*

Gelatine can be removed (e.g. by sodium hydroxide or enzyme digestion) from autoradiograms (Przybylski, 1961; Hay and Revel, 1963); this procedure should seldom be necessary and if employed subjects the autogram to the possibility of relocation or removal of silver grains.

6.15. *Improved Methods*

If the worker has by this stage obtained encouraging results, he may wish to improve on them. In particular, better resolution can be achieved using finer grained emulsions, such as Kodak NTE, and more subtle developing procedures. Complications will ensue with these fine grain emulsions, and SALPETER and BACHMANN (1964) in a very important paper describe various problems encountered and ways to circumvent them. These workers also give an excellent coating technique (see also SALPETER and BACHMANN, 1965). However, the research worker would possibly be advised to wait until he is reasonably experienced in autoradiography before attempting these.

In particular, the removal of gelatine from NTE emulsion prior to making the films is more difficult in my experience, than is indicated by SALPETER and BACHMANN (1964), a conclusion supported by GRANBOULAN (1965); their methods of gold latentisification of the image is difficult (rarely working in my inept hands!), and their coating method usually necessitates all batches of sections mounted on one slide being developed together (with the agar block method single grids can easily be developed at intervals to test exposure). However, the autoradiograms produced by these workers are superb, setting a standard which is at present unsurpassed. Using their methods of film preparation and physical development (see also CARO and VAN TUBERGEN, 1962), the smallest developed silver grains are about 500 Å across (*v.* 2000—4000 Å for Ilford L4 emulsion).

Improvement of sensitivity can be achieved by gold latentisification (SALPETER and BACHMANN, 1964) and possibly by exposure of the emulsion in a magnetic field (HARFORD and HAMLIN, 1961; NAKAI, 1964) although ROGERS (1967: p. 301) doubts the efficacy of this treatment. Scintillation autoradiography has also been suggested by PRZYBYLSKI (1969). KOPRIWA (1967a) has described a semiautomatic instrument for coating sections, later also publishing a detailed account of the characteristics of various developers when used with L4, NTE and Gevaert 307 emulsions and recommending the latter (KOPRIWA, 1967b). An excellent statistical approach to the problem of resolution has been given by SALPETER, BACHMANN and SALPETER (1969). BACHMANN and SALPETER (1967) describe a calibration method for measuring the absolute sensitivity of autoradiography under set conditions and the development of quantitative electron microscopy can be illustrated by SALPETER's (1967, 1969) work. FLITNEY (1969) describes an ingenious method for including an internal standard of radioactivity (as tritiated Araldite) in autograms.

6.16. *Some Problems of Interpretation*

In view of the often optimistic and naive expectations accompanying autoradiographic experiments alluded to earlier (Section 6.1), some important and serious drawbacks of autoradiography will be briefly mentioned for consideration.

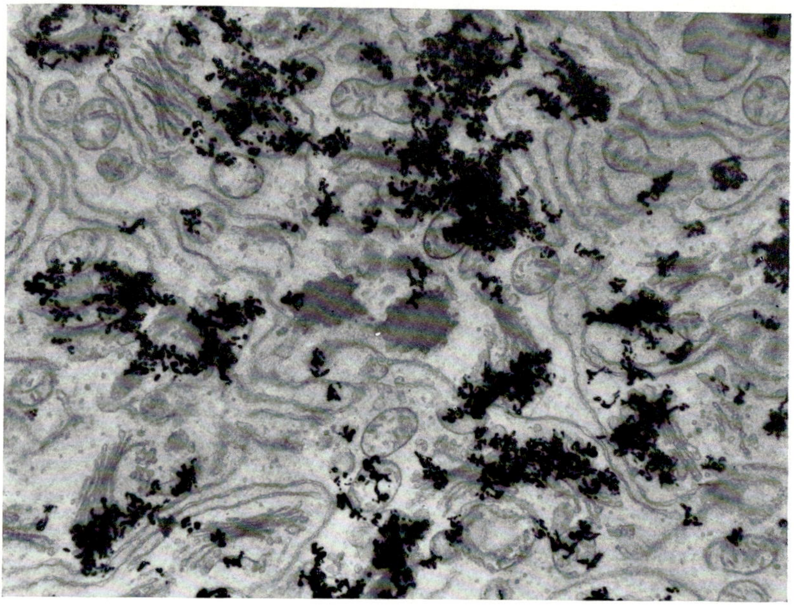

Fig. 6.5. Outer root cap cell, root tip exposed to glucose-6-H^3 for 10 minutes and then fixed in permanganate. Radioactivity is quite clearly very concentrated over golgi vesicles, although individual silver grains are frequent elsewhere. × 16,000 approximately

6.16.1. — Resolution

Autoradiography must be recognised as a statistical phenomenon. Even under optimal conditions, a "point source" of radioactivity may expose silver grains some distance from itself (see CARO, 1962; SALPETER, BACHMANN and SALPETER, 1969). So inferences drawn from selected high magnification autoradiograms showing isolated silver grains over cytoplasmic structures are risky unless further evidence confirms that these selected micrographs and their silver grain(s) represent fair statistical samples. An accompanying low power autoradiograph is generally valuable if it demonstrates what proportion of silver grains is localised over a specific cell structure. For example, in labeled root cap cells (Fig. 6.5), radioactivity is clearly concentrated in golgi bodies, even though many individual silver grains are seen elsewhere. In Fig. 6.6, the concentration of radioactivity in a section through an epidermal cell plate coincides very closely with the region occupied by cell plate vesicles, strongly supporting the inference that very small vesicles derived from the golgi bodies are themselves labeled (Fig. 6.7).

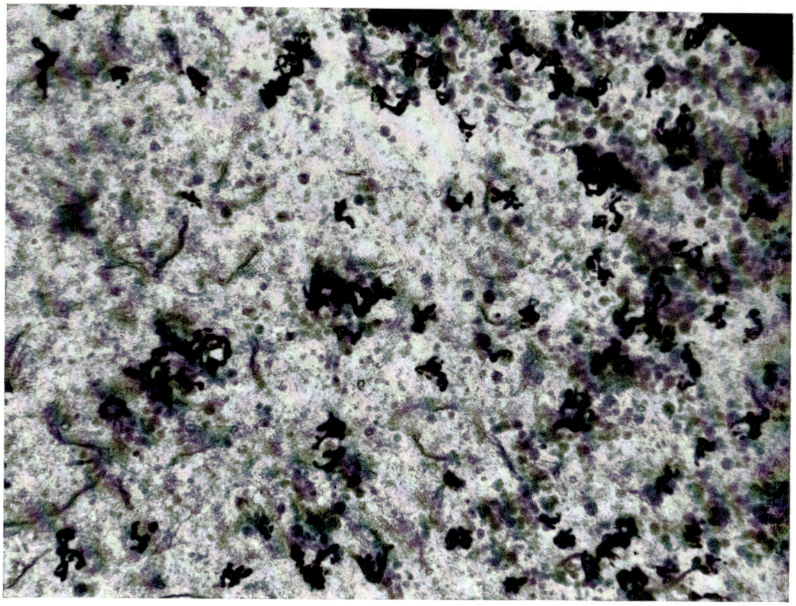

Fig. 6.6. Root-tip exposed to glucose-6-H^3 for ½ hour and then fixed in permanganate. This section, through and in the plane of an epidermal cell plate, shows radioactivity concentrated in the region occupied by the small cell-plate vesicles. This strongly supports the inference (e.g. in Fig. 6.7) that the golgi vesicles are labeled — even though the vesicles are below the limit of resolution achieved by method as applied here. × 14,000 approximately

RAY's (1967) excellent paper demonstrates how statistical autoradiography can be attempted on plant tissue. He studied the incorporation of tritiated glucose into the thick epidermal walls of oat coleoptiles and pea stems. By treating some of these tissue blocks with certain polysaccharide-extracting agents before embedding them, he was able to show that some labeled matrix polysaccharide material had moved a considerable distance through the cell wall after synthesis; this conclusions was based on a statistical comparison of the distribution of silver grains over the width of treated and untreated walls. The effect of growth hormone (IAA) on the incorporation of glucose into these polysaccharides was also investigated.

6.16.2. — Specificity

Certain autoradiographic results have disturbing implications, since even very specific precursors may not give the equivalent specific incorporations expected. Following thymidine incorporation in root tips, FUSSEL (1968)

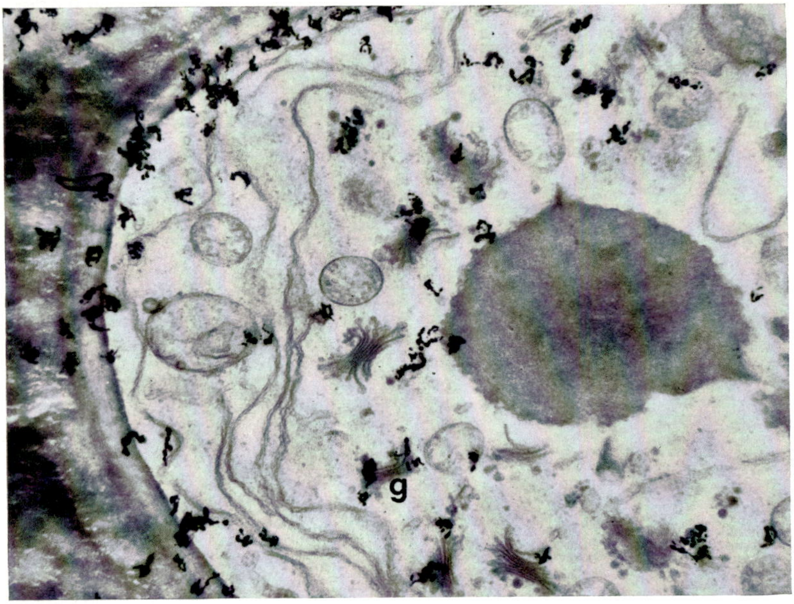

Fig. 6.7. Epidermal root tip cell, exposed to glucose-6-H^3 for 10 minutes and then fixed in permanganate. Cytoplasmic silver grains are concentrated over the golgi bodies (g) and most of this radioactivity is probably contained within the small golgi vesicles (c.f. Fig. 6.6), which are below the limits of autographic resolution achieved. × 10,000 approximately

found of course nuclei heavily labeled, but she also always detected some incorporation into the cytoplasm generally (and even into xylem wall thickenings); most significantly, she was unable by various chemical treatments (e.g. enzyme digestion) to characterise an appreciable, non-extractable fraction of these radioactive derivatives of thymidine. Her important conclusion is that until this fraction is identified, one is *not* justified in assuming that all the silver grains in such autoradiograms represent an incorporation of thymidine into DNA, an interpretation prevalent in the literature (see also Lima-de-Faria and Moses, 1964; Sagan, 1965; Schultze, 1969: p. 77 *et seq*.). This means (as before) that select micrographs containing isolated silver grains may by themselves illuminate little except the optimism of the author concerned. My own work with cinnamic acid (Pickett-Heaps, 1968) was based on biochemical considerations indicating that this precursor must go into lignin virtually exclusively; the results indeed showed massive incorporation into xylem wall thickenings as hoped (Fig. 6.8). However in this experiment, isolated silver grains were also found over all organelles, and — *reductio ad absurdum* — carefully selected micrographs could be said to "demonstrate" that any of these organelles (e.g. mitochondria, nucleoli etc.) were involved in lignin biosynthesis! The moral need not be stressed further.

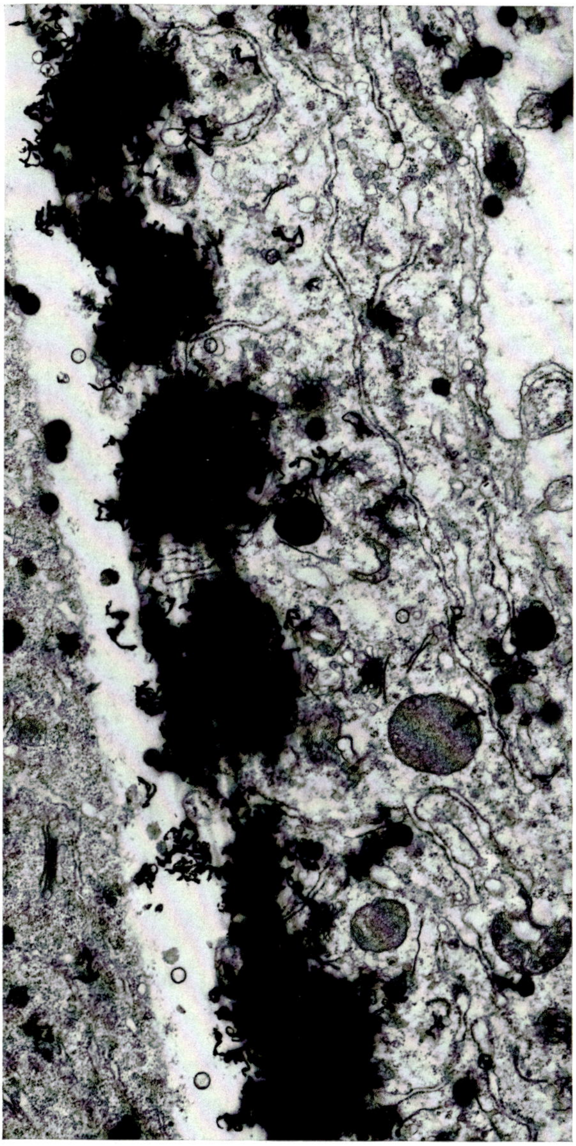

Fig. 6.8. Wheat coleoptile xylem cell; tissue exposed to tritiated cinnamic acid for 2 hours before being fixed in glutaraldehyde/osmium. The concentration of this highly specific lignin precursor in xylem wall thickenings is very impressive; however, individual silver grains can be found in such cells over all cell organelles if sought assiduously. Such individual silver grains prove nothing. × 14,000 approximately

Such results can be explained in diverse ways (e.g. enzyme-mediated exchange reactions involving the tritium atom, movement of labeled intermediates after fixation etc.); what is more important at the outset is to recognise this problem's existence. With luck, it may not appear, but in fact most published autoradiograms show this lack of specificity in organelle labeling to a widely varying extent. Attempts to demonstrate labeling of cell organelles must have some statistical basis — most easily provided by representative low power autograms from which one can estimate visually the specificity obtained.

6.16.3. — Pulse Experiments

It is worth pointing out one hazard in the interpretation of pulse experiments. Say the cell has two quite separate biochemical pools, A and B, utilising a common precursor, and that A is more rapidly turning over than B. During a short exposure to that labeled precursor, the organelles concerned with pool A will be more rapidly labeled than those with pool B; follow this short pulse with a cold chase and pool A's radioactivity will decrease whilst B's is still increasing. This may appear to demonstrate a passage of radioactivity from pool A to pool B, a problem difficult to eliminate. An irrefutable demonstration of such a passage of labeled material from one organelle or cell region into another is really only possible following the transplant of the labeled organelle into an unlabeled cell (achieved, for example, in demonstrating the passage of labeled RNA from the nucleolus into the cytoplasm — Goldstein and Plaut, 1955). Needless to say, in most experiments this procedure is at present quite impossible.

References

Albersheim, P., Neukom, H., Deuel, H.: Splitting of pectin chain molecules in neutral solutions. Archs. Biochem. **90**, 46—51 (1960).

Bachmann, L., Salpeter, M. M.: Absolute sensitivity of electron microscopic radioautography. J. Cell Biol. **33**, 299—305 (1967).

Bergeron, M., Droz, B.: Analysé critique des conditions de fixation et de préparation des tissus pour la détection radioautographique des protéins néoformées, en microscopie électronique. J. Microscopie. **7**, 51—62 (1968).

Caro, L. G.: High-resolution autoradiography II. The problem of resolution. J. Cell Biol. **15**, 189—199 (1962).

Caro, L. G.: A common source of difficulty in high-resolution radioautography. J. Cell Biol. **41**, 918—919 (1969).

Caro, L. G., Schnos, M.: Tritium and Phosphorous-32 in high-resolution autoradiography. Science **149**, 60—62 (1965).

CARO, L. G., TUBERGEN, R. P. VAN: High-resolution autoradiography I. Methods. J. Cell Biol. **15**, 173—188 (1962).

COPE, G. H., WILLIAMS, M. A.: Quantitative studies on neutral lipid preservation in electron microscopy. J. Roy. Micr. Soc. **88**, 259—277 (1968).

CRONSHAW, J., ANDERSON, R.: Sieve plate pores of *Nicotiana*. J. Ultrastruct. Res. **27**, 134—148 (1969).

DERMER, G. B.: An autographic and biochemical study of oleic acid absorption by intestinal slices including determinations of lipid loss during preparation for electron microscopy. J. Ultrastruct. Res. **22**, 312—325 (1968).

FEWER, D., THREADGOLD, J., SHELDON, H.: Studies on cartilage V. Electron microscope observations on the autographic localization of S^{35} in cells and matrix. J. Ultrastruct. Res. **11**, 166—172 (1964).

FLITNEY, F. W.: Tritium-labelled Araldite as an internal standard for quantitative autoradiography using the electron microscope. J. Roy. Micr. Soc. **90**, 151—156 (1969).

FUSSELL, C. P.: Cytoplasmic label following tritiated thymidine treatment of *Allium cepa* L. roots. J. Cell Biol. **39**, 264—285 (1968).

GOLDSTEIN, L., PLAUT, W.: Direct evidence for nuclear synthesis of cytoplasmic Ribose Nucleic Acid. Proc. Nat. Acad. Sci. U.S.A. **41**, 874—880 (1955).

GRANBOULAN, P.: Comparison of emulsions and techniques in electron microscopic radioautography. In: "The use of radioautography in investigating protein synthesis" (Symp. Int. Soc. Cell Biol. **4**), edited by C. P. LEBLOND and K. B. WARREN: Acad. Press, New York and London; 43—64 (1965).

HARFORD, C. G., HAMLIN, A.: A method of electron microscopic autoradiography with tritium-labelled cells in a magnetic field. Lab. Invest. **10**, 627—635 (1961).

HAY, E. D., REVEL, J. P.: The fine structure of the DNP component of the nucleus. J. Cell Biol. **16**, 29—51 (1963).

KIDWAI, P., ROBARDS, A. W.: The appearance of differentiating vascular cells after fixation in different solutions. J. exp. Bot. **20**, 664—670 (1969).

KOEHLER, J. K., MÜHLETHALER, K., FREY-WYSSLING, A.: Electron microscopic autoradiography. J. Cell Biol. **16**, 73—80 (1963).

KOPRIWA, B. M.: A semiautomatic instrument for the radioautographic coating technique. J. Histochem. Cytochem. **14**, 923—928 (1967a).

KOPRIWA, B. M.: The influence of development on the number and appearance of silver grains in electron microscopic radioautography. J. Histochem. Cytochem. **15**, 501—515 (1967b).

LIMA-DE-FARIA, A., MOSES, M. J.: Labeling of *Zea mays* chloroplasts with H^3-thymidine. Hereditas **52**, 367—378 (1964).

MARCHANT, H. J., PICKETT-HEAPS, J. D.: Ultrastructure and differentiation of *Hydrodictyon* I. Anst. T. Biol. Sci **23**, 1173—1186 (1970).

MAUNSBACH, A. B.: Absorption of I^{125}-labeled homologous albumin by rat kidney proximal tubule cells. J. Ultrastruct. Res. **15**, 197—241 (1966).

NAKAI, T.: A study of the ultrastructural localization of hair keratin synthesis utilising electron microscopic autoradiography in a magnetic field. J. Cell Biol. **21**, 63—73 (1964).

NORTHCOTE, D. H., PICKETT-HEAPS, J. D.: A function of the golgi apparatus in plysaccharide synthesis and transport in the root-cap cells of wheat. Biochem. J. **98**, 159—167 (1966).

NORTHCOTE, D. H., WOODING, F. B. P.: Development of sieve tubes in *Acer pseudoplatanus*. Proc. Roy. Soc. B. **163**, 524—537 (1966).

ORLIC, D.: The use of Fe^{55} in high-resolution radioautography of developing red cells. J. Cell Biol. **39**, 201—208 (1968).

Peters, T., Ashley, C. A.: An artefact in radioautography due to binding of free amino-acids to tissues by fixatives. J. Cell Biol. **33**, 53—60 (1967).

Pickett-Heaps, J. D.: Incorporation of radioactivity into wheat sylem walls. Planta **71**, 1—14 (1966).

Pickett-Heaps, J. D.: The use of radioautography for investigating wall secretion in plant cells. Protoplasma **64**, 49—66 (1967).

Pickett-Heaps, J. D.: Xylem wall deposition; radioautographic investigations using lignin precursors. Protoplasma **65**, 181—205 (1968).

Pickett-Heaps, J. D., Fowke, L. C.: Cell divison in *Oedogonium* I. Aust. J. biol. Sci. **22**, 857—894 (1969).

Przybylski, R. J.: Electron microscope autoradiography. Expl. Cell Res. **24**, 181—184 (1961).

Przybylski, R. J.: Scintillation radioautography: a new technique designed to augment silver grain number in radioautographs. J. Cell Biol. **43**, 108A (1969).

Ray, P. M.: Radioautographic study of cell wall deposition in growing plant cells. J. Cell Biol. **35**, 659—674 (1967).

Rogers, A. W.: Techniques of autoradiography. Amsterdam, London, New York: Elsevier Publishing Co. 1967.

Sagan, L.: An unusual pattern of tritiated thymidine incorporation in *Euglena*. J. Protozool. **12**, 105—109 (1965).

Salpeter, M. M.: Electron microscope radioautography as a quantitative tool in enzyme cytochemistry. I. J. Cell Biol. **32**, 379—389 (1967).

Salpeter, M. M.: Electron microscope radioautography as a quantitative tool in enzyme cytochemistry II. J. Cell Biol. **42**, 122—134 (1969).

Salpeter, M. M., Bachmann, L.: Autoradiography with the electron microscope. J. Cell Biol. **22**, 469—477 (1964).

Salpeter, M. M., Bachmann, L.: Assessment of technical steps in electron microscope autoradiography. In: "The use of radioautography in unvestigating protein synthesis" (Symp. Int. Soc. Cell Biol. **4**), edited by C. P. Leblond and K. B. Warren: Acad. Press, New York and London; 23—42 (1965).

Salpeter, M. M., Bachmann, L., Salpeter, E. E.: Resolution in electron microscope radioautography. J. Cell Biol. **41**, 1—20 (1969).

Schultze, B.: Physical techniques in biological research (edited by A. W. Pollister), Vol. IIIB. Acad. Press, London and New York.

Spurr, A. R.: A low-viscosity epoxy resin embedding medium for electron microscopy. J. Ultrastruct. Res. **26**, 31—43 (1969).

Stein, O., Stein, Y., Goodman, D. S., Fidge, N. H.: The metabolism of chylomicron cholesteryl ester in rat liver. J. Cell Biol. **43**, 410—431 (1969).

Wooding, F. B. P.: Radioautographic and chemical studies of incorporation into sycamore vascular tissue walls. J. Cell Sci. **3**, 71—80 (1968).

Chapter 7

Electron Probe Analysis[1]

ANDRÉ LÄUCHLI*
Department of Plant Sciences, Texas A & M University
College Station, Texas 77843, U.S.A.

Contents

[1] A contribution of the Texas Agricultural Experiment Station.

* Fachbereich Biologie-Botanik, Technische Hochschule Darmstadt, 6100 Darmstadt, Schnittspahnstr. 3—5.

7.1. *Introduction*

Progress in modern plant physiology has been largely dependent upon the development of new analytical methods. Each increase in analytical sensitivity has been accompanied by a wave of discoveries which were not observable before the new method was devised. To gain a deeper insight into the secrets of the plant's life, microanalytical methods are being employed through which the complex physiological and biochemical processes may be studied at the molecular level. In simple terms — what one really wants is to learn more and more about less and less. That the use of radioisotopes in autoradiography at the light and electron microscope level did contribute effectively to such progress has been demonstrated in the preceding chapters of this book.

Clearly, radioisotopes applied as tracers revolutionized methodology in physiological investigations of plants. There is a need, however, for also detecting chemical elements in tissues and cells. A method, with which this aim is to be achieved, not only must exhibit a detection sensitivity comparable to that of radioassays but also must be applicable to microscopical areas of the specimen. Electron probe analysis (EPA[2]) fulfills both demands.

Electron probe analysis is a method which permits an essentially complete elemental analysis of microscopically small regions in tissue sections. Its sensitivity is superior to that of chemical and spectroscopical methods by several orders of magnitude, the detection limits being approximately 10^{-15} to 10^{-16} g (Table 7.1). This compares favorably with the very sensitive radioassay of ^{32}P. Table 7.1 shows further that in terms of sensitivity

Table 7.1. Sensitivity Limits of Microanalytical Methods

Method	Sensitivity (g)	Reference
Chemical	10^{-9}	Marton (1969)
Absorption spectroscopy	10^{-9}	Marton (1969)
Laser probe analysis	10^{-11} (Mg)	Rosan, Brech and Glick (1965)
Emission spectroscopy	10^{-12}	Marton (1969)
Radioassay of ^{14}C	10^{-12}	Chase and Rabinowitch (1962)
Mass spectroscopy	10^{-13}	Marton (1969)
Electron probe analysis	10^{-15} to 10^{-16}	Andersen (1967)
Radioassay of ^{32}P	10^{-16}	Chase and Rabinowitch (1962)
Analytical electron microscopy (EMMA 4)	10^{-17}	AEI Scientific Apparatus Inc. (1970)
Ion probe mass analysis	10^{-20} (Na)	Galle, Blaise and Slodzian (1969)

[2] Abbreviation: EPA: electron probe analysis.

only two other methods may substitute for EPA; they are analytical electron microscopy and ion probe mass analysis. These are recent advances in methodology which will be described briefly in the section concerning "Related Methods" at the end of this chapter.

The scope of the present article is to provide the reader with the basic knowledge of what EPA is and how it may become a powerful tool in the hands of a plant physiologist whose prime interest is in the area of histological and cytological localization of chemical elements.

7.2. *Historical Remarks*

The concept of electron probe analysis (EPA) arose in the late 1940's from the synthesis of electron microscopy and x-ray spectrochemical analysis (BIRKS, 1963). CASTAING, while a graduate student of Professor GUINIER at Paris, France, designed the first instrument and established the theoretical background for quantitative analysis. This pioneering work is compiled in his dissertation (CASTAING, 1951). Similar electron probe analyzers were built thereafter in the United States and Russia, and soon were applied to the study of metallurgical and mineralogical problems.

ENGSTRÖM (1953), in a review on "X-Ray Methods in Histochemistry", was the first to mention the possibility of applying EPA to histochemical investigations of biological specimens. Four years later this became a reality when TOUSIMIS used the electron probe analyzer to examine the distribution of Ca in a section, 40 μ thick, of a human bone (cf. TOUSIMIS, BROOKS and BIRKS, 1959). This initial study stimulated many investigations of animal and human tissues in which EPA was utilized. Several reviews reported about the rapid progress thus made (TOUSIMIS, 1963a, b, c, 1966; HALL, HALE and SWITSUR, 1966; ANDERSEN, 1967; HALL, 1968a; HALL and HÖHLING, 1969; HÖHLING and HALL, 1969; TOUSIMIS, 1969a). Not only did EPA become an important method in animal and human biology, but advances were also made in instrumentation (CASTAING, 1960; MCKINLEY, HEINRICH and WITTRY, 1966; HEINRICH, 1968; TOUSIMIS and MARTON, 1969).

The first report on the cellular localization of mineral nutrients in plant tissues, using EPA, appeared in 1966 (LÄUCHLI and SCHWANDER, 1966). Since that time, plant physiologists became increasingly aware of the potentials of this new method. This is reflected in two early reviews (LÄUCHLI, 1968a, RASMUSSEN, SHULL and DRYER, 1968) and many recent original papers. Parallel to this development, EPA also proved to have broad applications to soil science (cf. CESCAS, TYNER and GRAY, 1968).

7.3. *Principles of Electron Probe Analysis*

7.3.1. General Description of the Method

In electron probe analysis (EPA), detection of chemical elements in tissues and cells is accomplished through identifying the characteristic x-ray spectra. Therefore, an electron probe analyzer consists of two principal components: a device which generates the characteristic x-ray spectra of the elements in the specimen, and a system which makes possible their qualitative and quantitative analysis. Generation of x-ray spectra is obtained in an electron optical system, and an x-ray detection system provides for analysis.

Fig. 7.1 represents a schematic drawing of the electron optical and x-ray detection systems. An accelerating potential between the filament of the gun and the anode, which usually can be varied from 0 to 30 KeV, accelerates electrons to move down the column toward the specimen at constant velocity. A vacuum of 10^{-5} Torr is maintained in the system. Size and current of the electron beam are controlled by the condenser lens while the objective lens focuses the beam onto the specimen. The nature of the objective lens determines the minimum beam diameter attainable (about 0.1 μ) and the maximum beam current, or in other words the spatial resolution of the method and its sensitivity respectively. The specimen is placed on a stage with which it is positioned relative to the beam. A reflected light microscope aids in bringing the specimen into position for analysis.

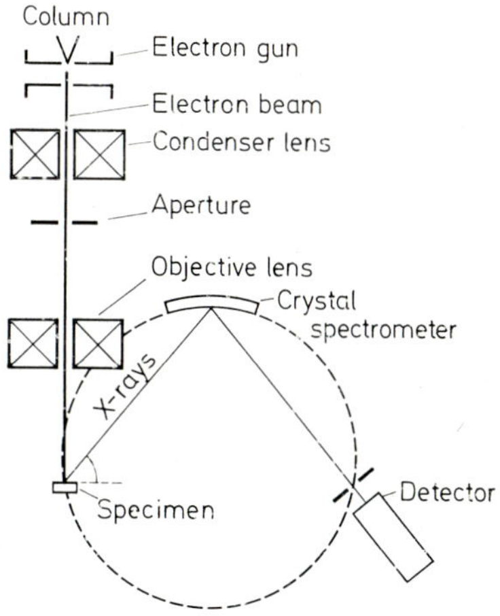

Fig. 7.1. Schematic drawing of the basic components of the electron probe analyzer

When the electron beam impinges upon the specimen, characteristic x-ray spectra are emitted from the excited chemical elements that are contained in the irradiated area. By chance, some of the high-energy electrons may collide with electrons in the K shell of atoms. When this occurs, the latter electron is ejected if the energy of the incoming electron is greater than the one of the K shell. The vacancy in the K shell will be filled by an electron from the L shell, the energy difference between the two electron shells being given off as an x-ray photon (in this case a (K_α) x-ray). If an L shell electron is ejected by a high-energy electron, the vacancy is filled by an electron from the M shell and an (L_α) x-ray is emitted, and so forth. The wavelengths of the characteristic x-rays increase from K to N shells of an element. Within the same electron shell, the wavelengths decrease with rising atomic number of the element (Table 7.2).

Table 7.2. Characteristic X-Rays of Some Biologically Important Elements: Detection with the Electron Probe Analyzer

Atomic number Z	Element	Critical excitation potential Kv (K_α) KeV[a]	Wavelength of (K_α) x-ray λ (Å)[a]	Analyzing crystal[b]	Wavelength range λ (Å)[b]
5	B	0.192	67	Pb-stearate	25.0 —93.0
6	C	0.283	44	decanoate	
7	N	0.399	31.6		
8	O	0.531	23.7	RAP[c]	6.57—24.76
11	Na	1.08	11.909	(RbH_2PO_4)	
12	Mg	1.303	9.889		
14	Si	1.837	7.126	ADP	2.66—10.00
15	P	2.142	6.155	($NH_4H_2PO_4$)	
16	S	2.470	5.373		
17	Cl	2.819	4.729		
19	K	3.606	3.744		
20	Ca	4.037	3.360	LiF	1.00— 3.77
26	Fe	7.109	1.937		
37	Rb	15.197	0.927	—	—
37	Rb	2.066 (L_α)	7.318 (L_α)	ADP	2.66—10.00
38	Sr	16.101	0.877	—	—
38	Sr	2.220 (L_α)	6.863 (L_α)	ADP	2.66—10.00

[a] Data from KEMP (1962).

[b] Data from KEIL (1967).

[c] RAP covers the same wavelength range as KAP (KH_2PO_4) but gives an approximately 8 to 10 times better peak to background ratio.

Elements are identified through wavelength and energy measurement by means of an x-ray detection system as shown schematically in Fig. 7.1. Wavelength measurement is accomplished by diffraction of the x-rays from a curved crystal in a crystal spectrometer. There are several crystals available, each of them covering a range of wavelengths and therefore suitable for analyzing a specific group of elements (Table 7.2). In addition to that, the analysis of a specific element also depends on the accelerating potential used because the latter must exceed the critical excitation potential for emitting a characteristic x-ray (see Table 7.2). A modern electron probe analyzer is equipped with an adjustable power supply so that the proper accelerating voltage can be selected and with several crystals so as to make possible the analysis of all elements heavier than Be.

An important feature in the performance of an electron probe analyzer is the x-ray take-off angle, i.e., the angle at which the crystal spectrometer senses the specimen (see Fig. 7.1). Low take-off angles lead to considerable absorption of x-rays in the specimen. This is an undesirable effect as sample topography influences x-ray emission, decreases sensitivity and makes quantitative analysis difficult.

After diffraction from a crystal the intensities of the characteristic lines are measured by a detector, usually a proportional counter (Fig. 7.1). A crystal spectrometer with a proportional counter form a dispersive detector. Two types of proportional counters are in use, the sealed and the gas-flow counters. The latter ones have extremely thin windows which are needed for detecting the long wavelengths of the characteristic lines of light elements with atomic numbers from 5 to 12. The proportional counter is not only a device for detecting intensities of characteristic lines but is also capable of determining their energies, for it produces a pulse, the amplitude of which is dependent upon the energy of the detected x-ray. Thus, energy measurement is obtained with a proportional counter in connection with a pulse height analyzer (Fig. 7.2).

One may analyze x-ray spectra through energy measurement without using diffracting crystals. This mode of analysis is known as energy dispersion (cf. Keil, 1967), accordingly the detector is of the nondispersive type. The energy resolution of a proportional counter extends only over about three atomic numbers, thereby limiting the utilization of the energy dispersion method if a single-channel pulse height analyzer is used. In contrast, energy dispersion with a multichannel analyzer offers a quick method for qualitative identification of elements in samples of unknown composition.

A new nondispersive detector, the lithium-drifted silicon detector, is now available from Ortec Inc., Oak Ridge, Tennessee 37830. Its characteristics have been described by Heinrich (1968). The advantage of the silicon detector as compared to the dispersive detector is its greatly in-

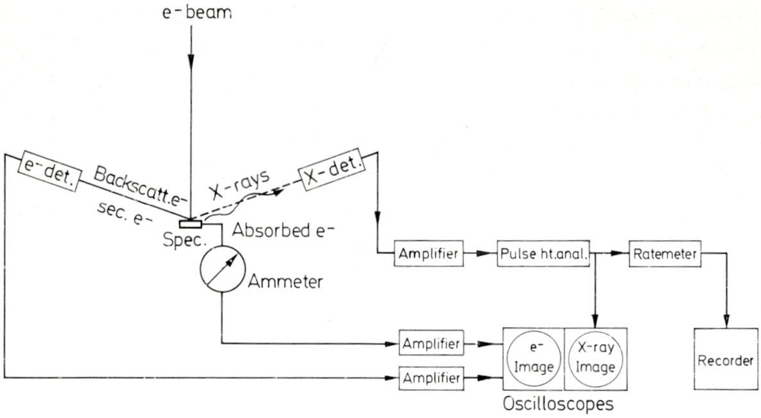

Fig. 7.2. Schematic diagram of the electron probe analyzer with scanning capabilities

creased speed. On the other hand, its peak to background ratio is poor which makes the silicon detector unfit to detect low elemental concentrations.

Thus far, the principles of the electron probe analyzer were described for the instrument in use with a static electron beam. A crucial step toward expanding the analytical capabilities was made when COSSLETT and DUNCUMB (1956) reported on "microanalysis by a flying-spot x-ray method". These investigators installed into the electron probe analyzer a device for scanning the beam across the specimen surface, i.e., they equipped an originally static instrument with scanning capabilities. In the scanning mode, the electron beam is deflected electrostatically or electromagnetically across the surface of the sample in a square raster. The information thus obtained is displayed on an oscilloscope whose flying spot sweeps in synchronization with the scanning electron beam. The x- and y-axes of the oscilloscope correspond exactly with the two dimensions of the scanned specimen area. Consequently, the oscilloscope display reflects a two-dimensional image of the specimen surface; for this reason, the method is named two-dimensional scanning.

Fig. 7.2 indicates that fourfold information is obtained from two-dimensional scanning. First, characteristic x-ray displays on the oscilloscope ("x-ray images") represent the pattern of elemental distribution in the scanned specimen area and may be looked upon as qualitative chemical maps. Second, three kinds of "electron images" can be displayed. The beam induces in the specimen an electrical current (= sample current) that passes through it and is measured with an ammeter. Then the current is amplified by a current amplifier, the output of which is used to modulate the intensity (z-axis) of the oscilloscope. Such an image is called "absorbed electron image". It conveys mainly information on the average mass density of the sample. The higher this density is, the more electrons are absorbed and the lower is the intensity of the oscilloscope image. The display of the "backscattered electron image" yields information inverse to that from the absorbed electron image in that the number of backscattered electrons rises with increasing mass density. Both absorbed and backscattered electron images give some information on the topography of the specimen. A very detailed image of the specimen's topography is obtained when the low-energy secondary electrons are collected and the "secondary electron image" displayed on the oscilloscope (DAVIDSON and NEUHAUS, 1966).

The utilization of a recorder connected with a ratemeter (Fig. 7.2) in wavelength and line scanning will be discussed in the section on "Analytical Techniques".

7.3.2. Analysis of Light Elements

Elements with atomic numbers from 5 to 12, which include such biologically important elements as C, N and O, emit (K_α) x-rays whose wavelengths are 10 Å or longer. Analysis of long wavelengths poses special problems, a full account of which has been given by ONG (1969). A dispersive detector suitable for measuring long wavelength x-rays consists of a crystal spectrometer with Pb-stearate decanoate and RAP-crystals (Table 7.2) and a gas-flow proportional counter.

In EPA, the beam cracks hydrocarbon molecules on the surface of the specimen. As a consequence a layer of carbon due to contamination builds up. This ought to be eliminated if C is to be analyzed. Carbon contamination also presents a hazard to the detection of other light elements since their soft x-rays may partly be absorbed by the contaminant. CAMPBELL and GIBBONS (1966) reported that the rate of carbon contamination in non-biological specimens varied from 1 to 10 Å/sec. With biological specimens, however, the contamination rate may be much faster and deposition of carbon on the surfaces of parts of the electron optical system may occur. Such contamination is reduced most efficiently by cooling the surrounding of the specimen which itself is kept at room temperature (CAMPBELL and GIBBONS, 1966; ONG, 1969). A cold plate around the specimen which is cooled with liquid N_2 may be utilized. Although such a cold trap virtually eliminates the problem of carbon contamination, it has the disadvantage of drastically reducing the space available to samples.

7.3.3. Spatial Resolution

One goal in EPA of biological specimens is to achieve a resolution on the order of 0.1 μ. Resolution depends on five factors: (i) diameter of the impinging electron beam, (ii) accelerating potential, (iii) specimen density, (iv) specimen thickness, (v) critical excitation potential. In instruments commercially available the lower limit of beam diameter is 0.1 μ. The actual resolution is reduced, for electron and x-ray scattering in the specimen lead to an increase in volume that contributes to the analysis and give rise to a secondarily excited area (Läuchli, Spurr and Wittkopp, 1970). The extent of this secondarily excited area is a function of accelerating potential and density and thickness of the specimen.

Fig. 7.3 shows a plot of the depth of electron penetration as a function of accelerating potential for two specimen densities, i.e., 0.6 and 1.0 g/ml respectively. The density of unembedded, freeze-dried plant material is estimated to be about 0.6 whereas that of resin embedded material is close to 1.0 (Läuchli, Spurr and Wittkopp, 1970). It follows from Fig. 7.3 that at high accelerating potentials the electron beam has a high penetration power in plant material. The secondarily excited area grows with increasing depth of electron penetration, hence the best resolution is obtained by analyzing thin specimens at low accelerating potentials (Andersen, 1967; Läuchli, Spurr and Wittkopp, 1970). Yet the critical excitation potential for generating the characteristic x-rays of each element (see

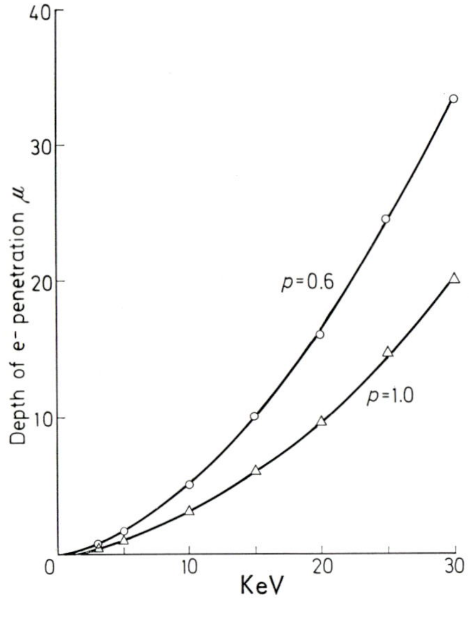

Fig. 7.3. Depth of electron penetration as a function of accelerating potential in unembedded, freeze-dried plant material (density $\varrho = 0.6$) and in resin embedded plant material (density $\varrho = 1.0$). (Data calculated from Andersen, 1967, Fig. 16)

Table 7.2) sets the lower limit of accelerating potential that can be used in the analysis. Further, the concentration of a particular element in plant material sometimes is not much higher than the detection sensitivity of the electron probe analyzer. The use of an accelerating potential only slightly above the critical excitation potential, at which the peak to background ratio is poor, is therefore not advised.

7.3.4. Detection Sensitivity

The absolute sensitivity limit in EPA is on the order of 10^{-15} to 10^{-16} g (see Table 7.1). Of great importance to the study of biological specimens is the relative sensitivity of the method in units of weight fraction. ANDERSEN (1967) defined the relative sensitivity as "the concentration of the element that is required to produce an x-ray intensity larger than the intensity of the background at the position of the analytical line by an amount equal to three times the standard deviation of the background intensity". For a specimen from a tooth, such relative sensitivity values as a function of accelerating potential have been determined by ANDERSEN and HASLER (1966) and are presented in Fig. 7.4. In general, sensitivities depend upon the accelerating potential used, and are usually best at two

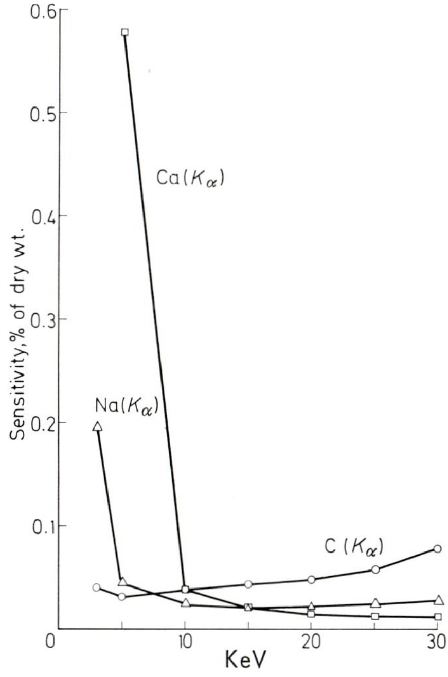

Fig. 7.4. Detection sensitivity of electron probe analysis as a function of accelerating potential for C, Na and Ca in a specimen from a tooth. Data are weight fractions, expressed in % of dry weight. The critical excitation potentials are 0.283 KeV for C(K_α), 1.08 KeV for Na(K_α) and 4.037 KeV for Ca(K_α). (From ANDERSEN and HASLER, 1966)

to six times the critical excitation potential. For analysis of C, having a critical excitation potential of only 0.283 KeV, maximum sensitivity is attained at 5 KeV. Consequently, light elements may be detected under conditions that permit maximum sensitivity and high spatial resolution. In contrast, heavier elements must be detected at higher accelerating potentials to obtain high sensitivity. Thus, the sensitivity limit for Na is 0.025% at 10 KeV and the limit for Ca, having a critical excitation potential of 4.037 KeV, is about 0.01% at 30 KeV (Fig. 7.4). It appears that for analysis of elements with atomic numbers between 12 (Mg) and 20 (Ca) one may compromise by choosing 15 KeV accelerating potential which assures reasonable sensitivity as well as fair resolution.

7.4. *Specimen Preparation*

Problems of specimen preparation in electron probe analysis (EPA) are similar to those in microautoradiography in so far as preservation of the chemical composition at the microscopic level is concerned. In addition to that, specimens to be used for EPA must fulfill specific requirements that are inherent to the method. Samples must be dehydrated completely before they are inserted into the vacuum of the instrument. Further, the samples preferably should be flat and exhibit a smooth surface, be electrically and thermally conductive, and be mounted on a suitable specimen support. The latter two factors will be discussed first since they can be treated equally with regard to the preparation of specimens for analysis of soluble and insoluble substances.

In choosing the material for specimen support, three factors must be taken into account, i.e., what elements are to be detected, the background radiation, and electrical and thermal conductivity. When thin specimens are used, which are completely penetrated by the electron beam, the chemical composition of the support must be chosen so as not to interfere with the analysis of the specimen. Since background radiation increases with rising atomic number of an element (HALL, HALE and SWITSUR, 1966; LÄUCHLI, SPURR and WITTKOPP, 1970) supports are being used that are made of light elements such as discs of Be, C or Al, quartz slides, and films of Formvar or Nylon. Thin plastic films appear to yield the least background radiation (HALL, HALE and SWITSUR, 1966) and are used routinely in HALL's laboratory to support animal specimens. The preparation of such Nylon films has been described by SIMS and HALL (1968). Commonly, a Nylon film of 0.2 μ thickness mounted on an Al tube, which is 1 cm in diameter, may support tissue sections up to several microns thick (HALL and HÖHLING, 1969).

LIBANATI and TANDLER (1969) have followed a similar approach for plant specimens by mounting 1 μ thick sections of corn root tips on Formvar-coated Cu grids.

Most studies on plant materials have been conducted with polished metallic supports of Al (LÄUCHLI, 1967a) or Be (LÄUCHLI, SPURR and WITTKOPP, 1970), or with polished C discs (RASMUSSEN, SHULL and DRYER, 1968). The lowest background is obtained with Be discs. However, Be is an expensive metal and may be replaced for economical reasons by C, the discs of which can be polished without difficulty in almost every laboratory. In contrast, the preparation of polished Al discs requires special equipment and skills (LÄUCHLI, 1967a), and attention must be paid to possible impurities of Mg.

Biological specimens are poor conductors of heat and electrical current. The impinging electron beam usually causes the temperature to rise considerably within a biological sample and renders the method destructive unless the sample is made conducting. Sufficient conductivity of unembedded materials is obtained by vacuum coating the samples on the support with approximately 200 Å C (LÄUCHLI, 1967a). However, it has been found necessary to coat epoxy resin embedded plant specimens with 150 to 200 Å Al for dissipation of heat (LÄUCHLI, SPURR and WITTKOPP, 1970). The electrical and thermal conductivity of plant specimens may be improved further by coating both sides of the samples with Al (cf. HALL, HALE and SWITSUR, 1966; LIBANATI and TANDLER, 1969). Yet one has to keep in mind that the background radiation rises substantially with increasing thickness of the Al coat, and the sensitivity for detecting light elements may drop because of x-ray absorption in the conducting coat.

7.4.1. Preparation Techniques for Detection of Soluble Substances

In preparing a specimen for analysis of soluble substances, it is essential to prevent during the entire procedure the displacement or leaching of the solutes to be detected. Unfortunately, there is probably no technique available which completely assures the achievement of this aim. Several procedures, however, have been worked out which minimize leaching to a large extent. LÜTTGE and WEIGL (1965) have quoted four techniques applicable to microautoradiography of soluble substances, i.e. (i) cryostat sectioning, (ii) freeze-drying and vacuum embedding in paraffin, (iii) freeze-drying, precipitation of the solute to be analyzed and paraffin embedding, and (iv) freeze-substitution and paraffin embedding.

7.4.1.1. Cryostat Technique

Most investigators consider the cryostat technique the safest method in so far as retention of soluble substances is concerned (Läuchli, 1966; Gahan et al., 1967; Rasmussen, Shull and Dryer, 1968; Rasmussen, unpublished results, 1969). In principle, the material is quickly frozen, the frozen tissue block is then transferred to the cryostat chamber, the temperature of which is usually maintained between —10 °C and —30 °C, and frozen sections are cut on the microtome installed within the cryostat chamber. The sections are placed on a specimen support and either allowed to dry in the chamber without thawing or freeze-dried. A detailed description of the technique follows.

After excision the tissue must be frozen rapidly. A matter of concern is whether or not a supporting medium should be used during freezing of the specimen. Gahan et al. (1967) froze the tissue on the block-holder without any supporting medium and thus avoided contamination of the specimen. However, subsequent cryostat sectioning of unsupported plant material often is difficult to accomplish and results in sections of poor quality. The specimen can be improved by mounting and freezing in a liquid supporting medium such as the Optimum Cutting-Temperature Compound OCT, manufactured by Ames Co., Elkhart, Indiana 46515 (Rasmussen, Shull and Dryer, 1968). If 5% propylene glycol is added to the OCT, the freezing point is lowered to —20 °C, and the specimen can be mounted in the still fluid supporting medium well below 0 °C (Satter, Marinoff and Galston, 1970). Yet, some leaching of soluble substances may occur during mounting of a specimen in OCT since it is a glycerol based compound. Exposure of the tissue to a liquid supporting medium is avoided if fresh brain tissue or other tissue from laboratory animals is used to support the specimens (Läuchli, 1966). Furthermore, the quality of the resulting sections is better as compared to those from unsupported specimens. Fresh animal tissues have been successfully employed in several laboratories to support plant specimens for EPA and microautoradiography (Crossett, 1967; Läuchli, 1967a; Waisel, Hoffen and Eshel, 1970). Clarkson and Sanderson (1970, see also Shone, Clarkson and Sanderson, 1969) have used pig's liver for supporting maize roots. These authors have demonstrated that a collar of an inert hydrocarbon mixture (i.e. either dodecane or tetradecane in olive oil) separating the roots from the liver will minimize the artificial redistribution of labeled ions due to ion exchange between the plant specimen and the supporting animal tissue.

There are several ways of adequately freezing the specimen for subsequent cryostat sectioning. The plant material may be directly frozen in a mixture of 8% (v/v) methylcyclohexane in isopentane cooled to —170 °C with liquid N_2, and then supported and mounted on the block-holder. Otherwise the tissue is supported and mounted on the block-holder and either frozen with burts of CO_2 (LÄUCHLI, 1967a) or frozen by immersing the metal base of the block-holder in isopentane cooled with liquid N_2 (CROSSETT, 1967). Thereafter, the block-holder is placed in the cryostat chamber and the specimen allowed to reach the temperature of the cryostat. Frozen sections are cut at temperatures in the range of —10 °C to —30 °C depending upon the physical characteristics of the specimen and of the supporting medium. Sections 10 to 12 μ thick are cut from meristematic tissues whereas the routine thickness to be obtained from differentiated plant material is 16 to 20 μ. With a fine brush the frozen sections are placed on the specimen support used in EPA, and spread by means of a drop of methylcyclohexane that is cooled to the temperature of the cryostat (LÄUCHLI and LÜTTGE, 1968). The sections are then allowed to dry in the cryostat chamber while still in the frozen stage, subsequently transferred on dry ice to an insulated box (e.g. made of styrofoam) which contains dry ice and a drying agent such as Drierite, and allowed to warm slowly to ambient temperature. Thus, the possibility of displacement of soluble substances due to water condensation is overcome (LÄUCHLI, 1966). Dehydration of the frozen sections is also accomplished through freeze-drying in a vacuum desiccator at about —20 °C (CHAYEN et al., 1960; PICKERING, 1966) or in a freeze-drier having a moving gas arrangement (BRANTON and JACOBSON, 1961).

In cryostat sectioning, attention has to be paid to several factors so as to minimize any secondary movement of substances during the actual preparation. First of all, the frozen block must not thaw nor may the sections thaw until they are entirely dehydrated. A crucial part of the procedure is the actual sectioning. If a dull knife is used the tissue is likely to thaw temporarily and rupturing may occur. A rapid cutting sequence may cause thawing as well.

Such an undesired effect, due probably to a dull knife, is demonstrated in Fig. 7.5. A transverse section, 20 μ thick, of a bean stem internode was cut on the cryostat and examined in the electron probe analyzer by two-dimensional scanning (RAINS and LÄUCHLI, unpublished data, 1969). Fig. 7.5a shows an absorbed electron image from an area depicting the transition zone of xylem (Xy) over cambium (Cm) to phloem (Ph). The knife apparently moved through the tissue perpendicular to the cambium and caused the phloem to rupture, leaving only scattered cell fragments. As a consequence, K is only localized in defined regions of the xylem but smeared over the area outside of the cambium (x-ray image of K (K_α) radiation in Fig. 7.5b). A mathematical evaluation of the probable degree of dislocation of materials in cryostat sections during a short term transient melting of the sections has been given by CLARKSON and SANDERSON (1970).

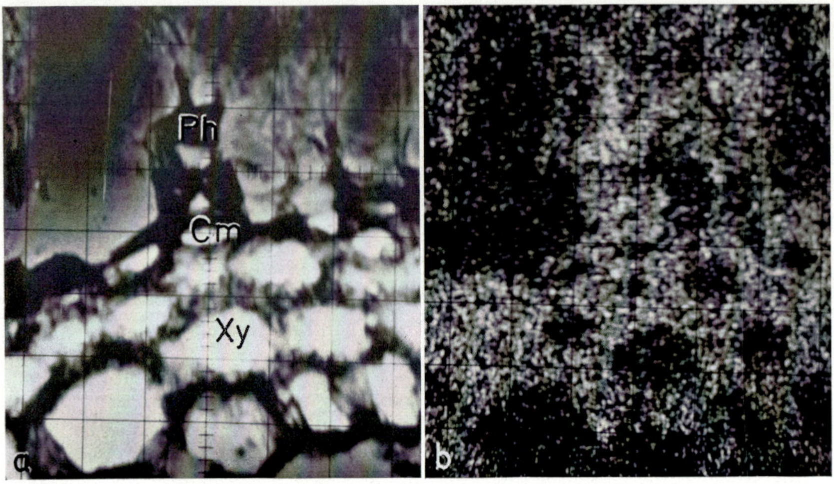

Fig. 7.5. Two-dimensional scan from a vascular bundle in a bean stem internode. Transverse cryostat section, 20 μ thick, mounted on an Al disc and coated with 150 Å C. a: absorbed electron image. b: x-ray image of K(K_α) radiation. Xylem (Xy), cambium (Cm), cell fragments of phloem (Ph). (RAINS and LÄUCHLI, unpublished data, 1969)

Most investigators who use cryostat sectioning of plant materials have experienced the frustrations of realizing that the preservation of cellular structures in the sections is poor. Further, cryostat sectioning of plant materials does not yield thin sections except for meristematic tissues (cf. KNOX, 1970). This is a limiting factor in achieving good spatial resolution in EPA (LÄUCHLI, SPURR and WITTKOPP, 1970). With animal materials, frozen sections thinner than 1 μ may be obtained at temperatures of about —70 °C to —150 °C (STUMPF and ROTH, 1965; APPLETON, 1968; HODSON and MARSHALL, 1970a). Recently, GAHAN, GREENOAK and JAMES (1970) described a method by which ultra-thin sections of unfixed frozen plant tissues may be prepared on a freezing ultramicrotome. It appears that the way is now open for the application of cryo-ultramicrotomy to studies of soluble substances by means of EPA.

7.4.1.2. Precipitation Techniques

Precipitation as a means for immobilizing solutes *in situ* is applicable only to those soluble substances that can be precipitated specifically and quantitatively to an insoluble product. Table 7.3 presents a survey of precipitation reactions in use to localize inorganic ions in biological materials. LÜTTGE and WEIGL (1962) have taken advantage of this possibility in a

Table 7.3. Precipitation Reactions for Detection of Inorganic Ions *in Situ*

Ion	Precipitating reagent	Method of localization	Reference
Na^+	K-pyroantimonate	TEM[a]	KOMNICK (1962), KOMNICK and KOMNICK (1963), BÖCK (1970), VAN LENNEP and KOMNICK (1971)
		TEM, EPA	LANE and MARTIN (1969)
Na^+	Mg-uranylacetate	TEM	KOMNICK (1962)
Na^+, Mg^{++} Ca^{++}	K-pyroantimonate	TEM, EPA	TANDLER, LIBANATI and SANCHIS (1970)
Ca^{++}	K-oxalate	Microautoradiography of ^{45}Ca	LÜTTGE and WEIGL (1962)
		TEM	CONSTANTIN, FRANZINI-ARMSTRONG and PODOLSKY (1965)
		EPA	PODOLSKY, HALL and HATCHETT (1970)
Cl^-	Ag-acetate or Ag-lactate	TEM	KOMNICK (1962), KOMNICK and KOMNICK (1963), ZIEGLER and LÜTTGE (1967), VAN LENNEP and KOMNICK (1971)
SO_4^{--}	Ba-acetate	Microautoradiography of $^{35}SO_4$	LÜTTGE and WEIGL (1962)
PO_4^{---}	Pb-acetate	TEM	TANDLER and SOLARI (1969)
		EPA	LIBANATI and TANDLER (1969)

[a] TEM: transmission electron microscopy.

microautoradiographic study on the distribution of $^{35}SO_4^{--}$ and $^{45}Ca^{++}$ in plant roots. Their procedure involved freezing followed by freeze-drying and then imbibing the tissue in aqueous solutions of either 5% Ba-acetate to precipitate SO_4^{--} or of 5% K-oxalate to immobilize Ca^{++}. Thereafter, the specimens were gradually dehydrated with ethanol and embedded in paraffin. In animal tissues, oxalate deposits have been demonstrated with the electron microscope (CONSTANTIN, FRANZINI-ARMSTRONG and PODOLSKY, 1965) and, subsequently, identified with EPA as Ca-oxalate crystals (PODOLSKY, HALL and HATCHETT, 1970).

A similar technique has been utilized to examine the distribution of PO_4^{---} ions (TANDLER and SOLARI, 1969; LIBANATI and TANDLER, 1969). Tissue segments were fixed in glutaraldehyde which also contained Pb-acetate to produce a Pb-orthophosphate precipitate *in situ*. Specimens were then dehydrated with ethanol and embedded in the plastic Maraglas. Thin sections were viewed with an electron microscope, while comparable sections, 1 μ thick, were analyzed for P and Pb with an electron probe analyzer. The monovalent ions Na^+ and Cl^- can also be localized intracellularly with transmission electron microscopy after precipitation of Na^+ with K-pyroantimonate or Mg-uranylacetate, and deposition of Cl^- with the acetate or lactate salts of Ag (ref. see Table 7.3). Owing to the great electron scattering power of the heavy elements Ag, Sb, Pb and U, such precipitates appear highly contrasted in the electron microscopic image.

The pattern of Cl distribution in *Limonium* leaves as revealed by electron microscopy of Ag^+-treated specimens is partly in accord with results from microautoradiographs of plant material that had absorbed ^{36}Cl (ZIEGLER and LÜTTGE, 1967). Moreover, microautoradiographic investigations of the transport of $^{35}SO_4^{--}$ and $^{45}Ca^{++}$ in plant roots led to similar results, whether the ions were precipitated as $BaSO_4$ and Ca-oxalate respectively or whether the cryostat technique was used for sample preparation (LÜTTGE and WEIGL, 1965). This suggests that precipitation techniques yield reliable results in localization experiments when such specific and straightforward reactions as those for deposition of Ca^{++}, Cl^- and SO_4^{--} are applied, though some intracellular displacement of these solutes is likely to take place during the formation of the precipitate. However, it is difficult to decide whether ionic distributions obtained through employing more complex reactions as for instance the immobilization of Na^+ with K-pyroantimonate coincide with the situation in the living tissue. On the one hand, LANE and MARTIN (1969) have shown with the electron probe analyzer that the pyroantimonate reaction does result in the specific deposition of Na salts in animal specimens. In contrast, recent evidence obtained with EPA by TANDLER, LIBANATI and SANCHIS (1970) indicates that K-pyroantimonate produces intracellular precipitates with Na^+ as well as with Mg^{++} and Ca^{++} in both animal and plant tissues. It appears that the results published thus far are not sufficient to assess the general validity of localizing inorganic solutes by precipitation.

7.4.1.3. Freeze-Drying and Vacuum-Embedding

The three main steps of this technique are freezing of the tissue, preferably at the temperature of liquid N_2, followed by freeze-drying, and finally embedding of the dehydrated tissue in paraffin under vacuum.

Standard paraffin sectioning is employed. BIDDULPH (1956) was the first to use the freeze-drying technique for microautoradiography of soluble substances in plant tissues.

Freeze-drying and embedding *in vacuo* is usually accomplished in a specially designed apparatus. Several types of such apparatuses have been described in the literature (BIDDULPH, 1956; BRANTON and JACOBSON, 1961; JENSEN, 1962; ESCHRICH, 1966). An extensive study on freeze-drying of plant materials was made by BRANTON and JACOBSON (1961) who achieved good results with a moving gas freeze-drying apparatus. In the laboratory of LÜTTGE (personal communication) an easy method for vacuum embedding of tissue was developed. The freeze-dried plant material was placed in test tubes on solid paraffin. The test tubes were evacuated by means of a Leybold-pump and transferred to a water-bath at 60 °C to melt the paraffin. With this technique embedding is completed in 5 to 7 days.

One difficulty of the freeze-drying technique lies in the fact that vacuum embedding in paraffin often is not completely successful. With regard to EPA, sample preparation is complicated further in that the paraffin sections must be dewaxed (with xylene) because paraffin does not withstand exposure to the electron beam and volatilizes. In special cases, one may elude these problems by examining unembedded, freeze-dried material with the electron probe analyzer. Such an approach has been used with freeze-dried portions of epidermal strips of leaves (SAWHNEY and ZELITCH, 1969).

Studies with the electron probe analyzer by ANDERSEN (1967) on amphibian red blood cells and unpublished data by RASMUSSEN (1969) from root and stem tissues of bean indicate that inorganic solutes like Na^+, K^+ and Cl^- may be lost to a considerable extent from the specimen if the freeze-drying technique is applied. Such a loss of solutes may occur during dewaxing with xylene, as LÜTTGE and WEIGL (1965) have shown in a microautoradiographic study, where dewaxing is not necessary, that labeled inorganic ions are retained in freeze-dried, vacuum embedded material. Additional studies are needed to resolve this question.

7.4.1.4. Freeze-Substitution

In addition to freeze-drying, dehydration of frozen plant specimens can also be achieved by freeze-substitution. JENSEN (1962) has described a procedure for removal of the ice from the tissue by dissolution in absolute ethanol at —30 °C to —40 °C. After dehydration the ethanol is replaced by toluene, the specimens brought to room temperature and embedded in paraffin. When the technique of freeze-substitution is used, difficulties in embedding are not encountered normally, as may be the case for freeze-dried material. However, those substances which are soluble in ethanol or toluene are lost from the tissue when JENSEN's method is employed. If Ca is to be localized in plant materials, its loss can be prevented by freeze-substitution with acetone instead of ethanol (GIELINK, SAUER and RINGOET, 1966).

For detection of soluble ions, LÜTTGE and WEIGL (1965) have worked out a freeze-substitution technique utilizing ether to remove the ice from the frozen specimens. This technique has been applied to microautoradiographic investigations of ion transport in plants (KRICHBAUM, LÜTTGE and WEIGL, 1967; ZIEGLER and LÜTTGE, 1967; OSMOND et al., 1969). The advantages appear to be a minimum loss of hydrophilic materials from the tissue in freeze-substitution, as a result of ether being very slightly miscible with water, and much less shrinkage or distortion of the tissue than when freeze-drying is being used. To assure the retention of solutes, ether must be dehydrated with Al_2O_3 prior to use, and Al_2O_3 must also be added to the ether during freeze-substitution for trapping the tissue water that is dissolved gradually by the ether (LÜTTGE and WEIGL, 1965; LÄUCHLI, SPURR and WITTKOPP, 1970).

Since paraffin embedded plant material is unfit for EPA unless the sections are dewaxed, LÄUCHLI, SPURR and WITTKOPP (1970) have worked out a new freeze-substitution, epoxy resin embedding technique involving anhydrous ether as the substituting fluid. The paramount feature of the new technique is the fact that thin sections, 1 μ thick, may be obtained anhydrously for high-resolution EPA. Furthermore, it allows retention of inorganic solutes at the original sites in the tissue and is superior to cryostat sectioning in the preservation of cellular structures. Resin embedded tissue bears still another benefit to EPA in that variations in specimen density are largely alleviated (see LÄUCHLI, SPURR and WITTKOPP, 1970 for a detailed description).

Briefly, tissue segments about 1 mm in length are cut (preferably in the cold, at 2 to 4 °C) and immediately frozen in a mixture of 8% (v/v) methylcyclohexane in isopentane cooled to —170 °C with liquid N_2. Freeze-substitution is accomplished using anhydrous ether at —30 °C. The dehydrated specimens are gradually infiltrated with Spurr's low-viscosity epoxy resin embedding medium E (SPURR, 1969) at temperatures below 0 °C. Final embedding and curing are done as described by SPURR (1969). Sections 1 to 2 μ thick are cut anhydrously with a diamond knife, using hexylene glycol in the ultramicrotome trough (cf. PEASE, 1966). The sections are finally mounted on the polished surface of a Be disc and vacuum coated with 150 to 200 Å Al.

Monitor castings for transmission electron microscopy may be obtained by a slightly altered procedure (see LÄUCHLI, SPURR and WITTKOPP, 1970; LÄUCHLI, SPURR and EPSTEIN, 1971).

Epoxy resins based on epichlorohydrin may contain small amounts of Cl (INGRAM and HOGBEN, 1968; LÄUCHLI, SPURR and WITTKOPP, 1970). The Cl content of Spurr's low-viscosity epoxy resin embedding medium E is about 0.073%. This level is too high for studies in which Cl is the element of interest, yet, it is possible to prepare epoxy resins with very low levels of Cl. Another alternative is to consider embedding in methacrylate (GIELINK, SAUER and RINGOET, 1966), but GALLE and STUVE (1968) do not recommend it because of its relative instability under the electron beam. Methacrylate also has other disadvantages such as marked shrinkage and bubble formation during polymerization.

Electron probe analysis of freeze-substituted, epoxy resin embedded tissue is the preferred method when soluble substances are to be analyzed on a cellular level, because it permits satisfactory preservation of cellular structures and a good spatial resolution in analysis. In contrast, cryostat sectioning is inferior with regard to the quality of the sections as well as resolution, and is advised as an alternative method only for studies on a histological level.

7.4.2. Preparation Techniques for Detection of Insoluble Substances

Thus far, techniques of specimen preparation have been described which permit the analysis of both soluble and insoluble substances. Electron probe analysis is also a tool for specific detection of deposits of insoluble minerals. In addition, possibilities exist for determining insoluble organic compounds. Preparation techniques for detection of insoluble substances will now be described separately.

Insoluble substances in animal tissues have been studied by TOUSIMIS (1963a, c). The material was usually fixed with formaldehyde, then dehydrated with ethanol or acetone and finally embedded in paraffin or some plastic material. For EPA of plant specimens RASMUSSEN (1968) and RASMUSSEN, SHULL and DRYER (1968) have applied two different preparation techniques in a study on the distribution of Al, Ca, P and K in corn seedlings. The cryostat technique as described earlier was used to analyze total contents of these elements while the standard paraffin method was employed to detect the insoluble fractions.

It appears that freeze-substitution, epoxy resin embedding or the cryostat technique are the methods of choice if the retention of both soluble and insoluble substances is desired. However, if only insoluble substances are to be detected, the most efficient procedure would seem to wash out specifically the soluble fraction and then prepare the specimens by means of a method that yields thin sections for attaining high resolution.

Specific washing procedures must be worked out individually for each particular study and no general recommendation can be made. As to the subsequent preparation of the specimen the following guidelines may be observed. Fix the material in a medium that preserves the cellular structures well, i.e., OsO_4, acrolein, glutaraldehyde, formaldehyde (FEDER and O'BRIEN, 1968). Dehydrate with a solvent compatible with the embedding medium to be used (cf. FEDER and O'BRIEN, 1968; SPURR, 1969), and embed in SPURR's low-viscosity embedding medium (SPURR, 1969). This medium has been found to withstand exposure to the beam in EPA (LÄUCHLI, SPURR and WITTKOPP, 1970).

There is still another approach to prepare specimens for EPA of insoluble minerals, that is microincineration. ALBRIGHT and LEWIS (1965) have detected Fe and Cu in microincinerated preparations of mineral-metabolizing bacteria and suggested that microincineration is advantageous to EPA in situations where the metal content is low. In microincineration, thin preparations of biological material, i.e., tissue sections or smears of cells or tissue, are mounted on a support (e.g. a quartz slide) and ashed. The organic material is oxidized completely and what remains is a pattern of the heat-stable mineral substances. Ashing may be achieved either at about 500 °C in a muffle furnace (high-temperature ashing) or by using low-temperature, active oxygen ashing, developed by GLEIT and HOLLAND (1962). In low-temperature ashing, temperatures below 100 °C are being maintained, assuring that metallic elements such as Na, K, Mg, Ca (SANUI and PACE, 1966), and phosphate (THOMAS and CORLETT, 1969) are quantitatively retained. The "state of the art" of microincineration of biological specimens has reached an advanced level in that it is now applied to electron microscopic studies, primarily through low-temperature ashing of samples mounted on grids of stainless steel (THOMAS, 1969).

Soluble mineral substances are likely to be displaced intracellularly in microincineration. Yet, in cases where microincineration is indicated (see THOMAS, 1969), the identification and localization of specific minerals in the ash may be accomplished most efficiently through EPA. This technique may even gain additional importance for analyzing very low concentrations of certain minerals which could not be detected in unashed specimens. THOMAS (1969) has projected the detection sensitivity for EPA of microincinerated specimens to be 10^{-17} or 10^{-18} g.

7.5. *Analytical Techniques*

Electron probe analysis (EPA) comprises a variety of modes of analytical operation yielding qualitative and quantitative data. The investigator will usually gather first some qualitative information on a specimen of unknown composition in order to obtain a survey of the total sample. He will then turn to quantitative detections in tissue regions that are of special interest for solving a particular problem.

7.5.1. Qualitative Analysis

7.5.1.1. Two-Dimensional Scanning

With the aid of the reflected light microscope the general area in the specimen, which is to be analyzed, is brought under the cross hairs. The electron beam scanning system is then operated so as to sweep the beam across the surface in a square raster, and electron and x-ray images are displayed on oscilloscopes. The image magnification is varied by changing the beam scanning deflection voltage, and covers in modern instruments the range from very low magnifications up to 10,000 times or higher. Electron and x-ray images displayed on the oscilloscopes may be photographed for permanent records.

Useful first information is obtained through the absorbed and backscattered electron images. Both displays reflect rough images of the specimen surface and of the average mass density or the average atomic number of the sample. They may be looked upon as very preliminary chemical maps. The great merits of displaying such electron images lie in the fact that areas of major inhomogeneities in the average atomic number are spotted easily. The absorbed electron image represents the difference in current between the impinging beam and the backscattered electrons. The intensity of the oscilloscope display therefore decreases with increasing average atomic number of the analyzed region. In contrast, the backscattered electron current is directly related to the average atomic number. A more detailed knowledge of the specimen surface is obtained by displaying the secondary electron image.

A specific chemical map is obtained for any given element through recording its characteristic x-rays on the oscilloscope. Such x-ray images even bear semiquantitative information, for areas of high elemental concentration produce more intense x-ray images than do areas of low concentration. Among all the possible analyses of characteristic x-rays with the electron probe analyzer, x-ray images obtained through two-dimensional scanning represent the least sensitive detection method.

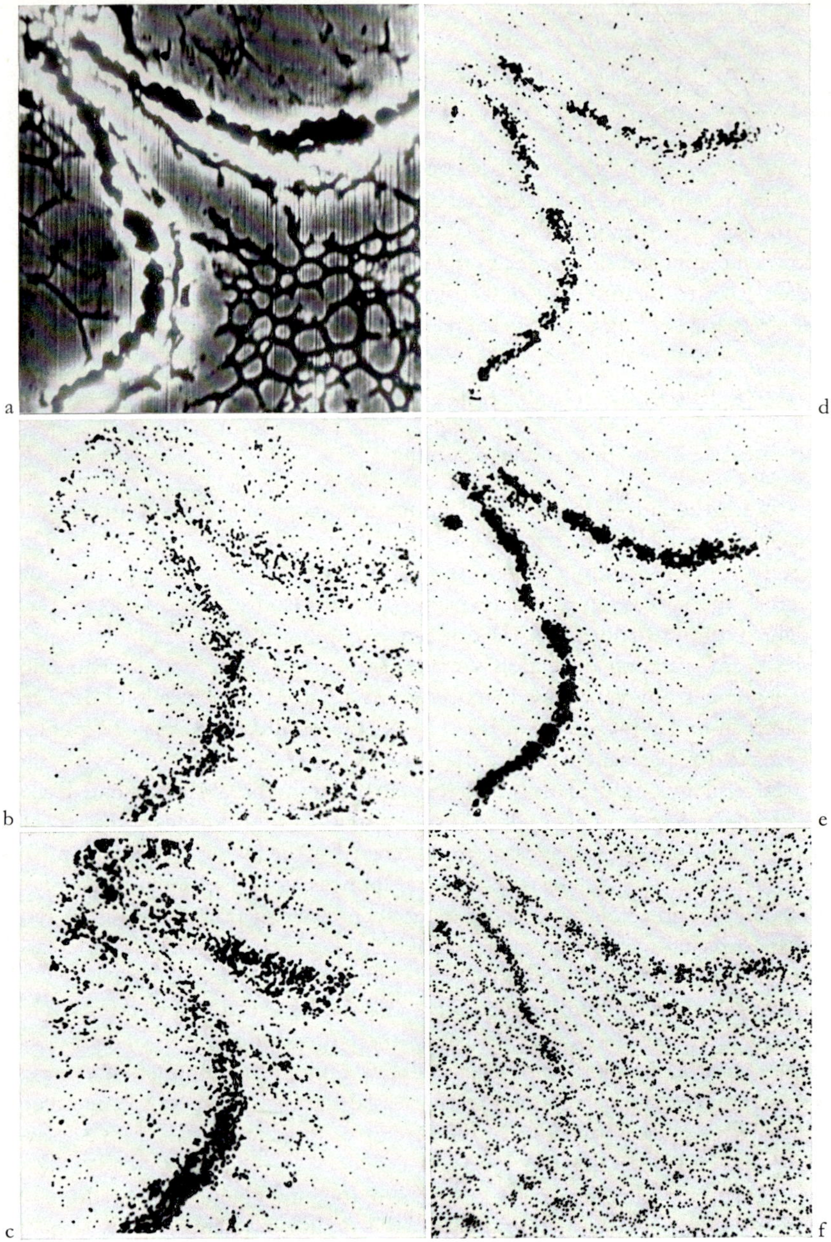

Fig. 7.6. Two-dimensional scan from an area depicting parts of the placenta and marginal sclerenchyma in the ventral suture of a pod of *Pisum sativum*. Transverse cryostat section, 20 μ thick, mounted on an Al disc and coated with 150 Å C. a: absorbed electron image. b to f: x-ray images of K(K_α), P(K_α), Ca(K_α), Sr(L_α), S(K_α) radiation respectively. × 300. (From LÄUCHLI, 1968b)

In Fig. 7.6 an example is given of two-dimensional scanning of a plant specimen. It is taken from a study on the translocation of inorganic solutes to the fruits of *Pisum sativum* (LÄUCHLI, 1968b) and exhibits the distribution of several elements in parts of the placenta and marginal sclerenchyma of the pod's ventral suture. Fig. 7.6a indicates that the intensity of the absorbed electron image is a function of both the topography of the specimen and its average atomic number. One may distinguish three phases in the sclerenchyma outside of the placental cells. The bright phase in the center of the sclerenchyma is due mainly to organic compounds whereas the two dark phases at the sclerenchyma's margins reflect inclusions of materials having a high average atomic number. The inner dark phase toward the placenta is caused primarily by high concentrations of K (Fig. 7.6b) and also of P (Fig. 7.6c). In contrast, deposits of Ca (Fig. 7.6d) and Sr (Fig. 7.6e) together with some accumulation of S (Fig. 7.6f) appear to have induced the outer dark phase.

7.5.1.2. Wavelength Scanning

Specific phases of microscopically small dimensions which become evident in absorbed or backscattered electron images may be analyzed qualitatively by wavelength scanning and a static electron beam. This is attained by scanning the x-ray spectrometers through their entire wavelength ranges and recording a complete x-ray spectrum with a strip chart recorder (cf. Fig. 7.2).

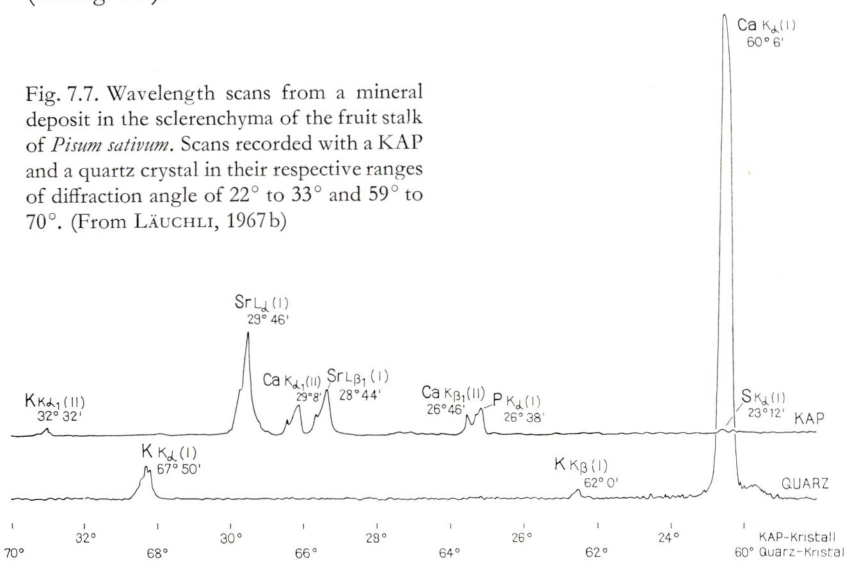

Fig. 7.7. Wavelength scans from a mineral deposit in the sclerenchyma of the fruit stalk of *Pisum sativum*. Scans recorded with a KAP and a quartz crystal in their respective ranges of diffraction angle of 22° to 33° and 59° to 70°. (From LÄUCHLI, 1967b)

A typical example of an x-ray spectrum is shown in Fig. 7.7. It reveals the elemental composition of a mineral deposit that was spotted in electron images from the sclerenchyma of a vascular bundle in the fruit stalk of pea (LÄUCHLI, 1967b). The two recorded traces represent in relative units parts of wavelength scans obtained with a KAP and a quartz crystal respectively. Also indicated are the respective ranges of diffraction angle that were covered (cf. BIRKS, 1963, Chapter 6, for a complete outline of x-ray crystal optics). Note the strong first-order maximum of Ca (K_α) radiation recorded with the quartz crystal at 60°6′ angle of diffraction as compared to the relatively weak second-order Ca (K_α) maximum at 29°8′ diffraction angle which was detected with the KAP crystal (the first-order is the principal diffraction while higher orders are considered as interferences). One may conclude from Fig. 7.7 that the mineral deposit consisted mainly of Ca-Sr-phosphate in a microcrystalline form.

7.5.2. Quantitative Analysis

7.5.2.1. Point Analysis

The most sensitive method of obtaining data in EPA is called point analysis. Specific specimen areas to be analyzed are brought under the cross hairs of the reflected light microscope. A static electron beam is then focused on the surface of the specimen. The characteristic x-ray spectra generated are analyzed by setting the x-ray spectrometers on the most sensitive lines of specific elements and recording the intensities of the peaks of the selected x-ray lines. X-ray intensities may be obtained manually from the scaler readouts or printed out with a typewriter. The readouts are in counts per unit time and represent a relative measure of the amount of an element per volume analyzed. The conversion of the relative quantitative data to absolute units will be discussed in Section 7.5.2.4. With regard to the statistical analysis of the quantitative data the reader is referred to ANDERSEN (1967).

At this point it is appropriate to mention that the operation of the electron probe analyzer may be computer controlled through utilizing a computer-probe-interface (BAYARD, 1969b). If there is a small computer available, an interface to be used as a data gathering system for the electron probe analyzer may be constructed with comparatively little investment. Such an interface is currently in operation for biological work (INGRAM, 1969).

The extent of the specimen area analyzed is directly related to the size of the impinging electron beam. It is customary to work with beams as small as possible having a diameter well below 1 μ to achieve good resolution. However, there are instances where integrated values are desired from larger areas of the specimen such as from a whole cell rather than a cell compartment. The width of the electron beam is then adjusted so as to cover completely the area of the cell. Employing large electron beams is also beneficial in K analyses because the volatilization of this element is thus eliminated (SATTER, MARINOFF and GALSTON, 1970). Further, the build-up of heat in biological specimens is distributed over a larger area and the heat effect by the beam is lessened.

7.5.2.2. Line Scanning

A quantitative picture of the pattern of distribution of a particular element is yielded by the line scan technique. A line scan is achieved on any line of traverse across the surface of a specimen when either the specimen is moved under the static electron beam by means of a stage drive motor or the electron beam is driven across the stationary specimen through operating the beam scanning system. In the former mode a permanent record of the line scan is obtained with the strip chart recorder. In contrast, the record of a line scan produced through the beam scanning system is acquired by photographing the oscilloscope display. Line scans yield quantitative data in relative units and reflect profiles of elemental concentration. When a series of such profiles is obtained in a kinetic study, line scanning produces information that bears a dynamic component. Thus, it is particularly useful in studying the pattern of translocation of inorganic solutes in plants and animals.

Fig. 7.8 presents the strip chart record of a line scan revealing the pattern of K distribution in a transverse section of a corn root. To provide high spatial resolution of the peaks of K(K_α) radiation, this particular line scan was obtained from a 2 μ thick section with the electron beam diameter being 0.5 μ and the stage motor being driven at a speed of 16 μ/min (cf. LÄUCHLI, SPURR and WITTKOPP, 1970). Thus, distinct peaks of K(K_α) radiation were obtained which can be assigned to specific types of cells such as the xylem parenchyma cells (LÄUCHLI, SPURR and WITTKOPP, 1970; LÄUCHLI, SPURR and EPSTEIN, 1971).

As in point analysis one may enlarge the beam diameter so as to achieve line scans that cover whole cells. Line scans having a width of several μ were used by HUMBLE and RASCHKE (1971) to study the accumulation of K in guard cells of leaves. This procedure reduced damage due to heat and abolished the volatilization of K which had been observed by SAWHNEY and ZELITCH (1969).

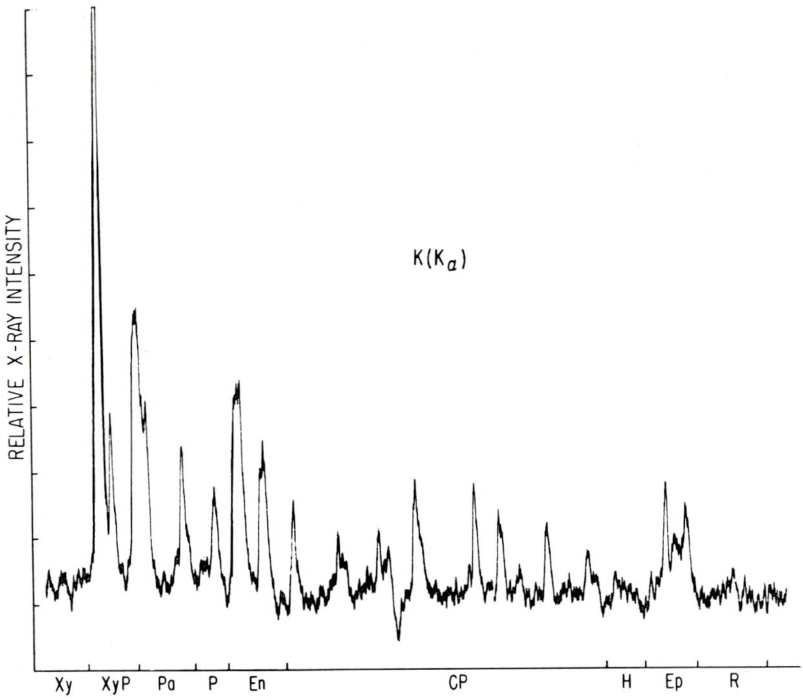

Fig. 7.8. Line scan of K(K_α) radiation across a 2 μ section of a corn root 10—11 mm behind the tip. Root segment frozen at —170 °C, freeze-substituted with anhydrous ether and embedded in low-viscosity epoxy resin medium. Section mounted on a Be disc and coated with 150 Å Al. Epoxy resin outside of tissue (R), epidermis (EP), hypodermis (H), cortical parenchyma (CP), endodermis (En), pericycle (P), unspecified stelar parenchymatous tissue (Pa), xylem parenchyma (XyP), xylem vessel (Xy). (From LÄUCHLI, SPURR and WITTKOPP, 1970)

7.5.2.3. Measurement of Total Mass and Elemental Weight Fractions

A method has been recently developed for measuring relative weight fractions of an element in thin biological specimens (HALL, 1968a; HALL and HÖHLING, 1969; HALL and WERBA, 1969). HALL's work was initiated partly because unembedded biological specimens exhibit an inhomogeneous distribution of dry mass per unit volume. This problem appears eliminated if the samples are embedded in a medium such as epoxy resin (LÄUCHLI, SPURR and WITTKOPP, 1970). However, for unembedded specimens, Hall's method represents a new approach which has been successfully applied to relatively thin animal specimens of a thickness on the order of 5 μ (SIMS and HALL, 1968). Although thin sections of unembedded plant material have not been yielded routinely with the techniques that are presently available (however see GAHAN, GREENOAK and JAMES, 1970), a brief outline of HALL's method will be presented to provide the background knowledge for possible applications to plant specimens in the future.

The new method is based upon the utilization of the intensity of white x-ray radiation ("Bremsstrahlung") as a direct measure of total mass per unit area. White x-ray radiation is observed by directly measuring a suitable band of quantum energies with a separate detector which does not have a crystal for x-ray diffraction. In a thin specimen of uniform thickness, mounted on a very thin support such as a film of Nylon, total mass per unit area is related to total mass per unit volume by a constant factor. Further, the intensity of the characteristic x-ray radiation of an element is proportional to the elemental mass per unit area. The ratio of the intensities of characteristic and white x-ray radiation is then a relative measure of the weight fraction (C) of the element (HALL, 1968a)

$$C = \frac{\text{mass of element}}{\text{total mass}} = \frac{\text{elemental mass/area}}{\text{total mass/area}} .$$

This equation is valid if the elemental mass is small compared to the total mass, and if the average atomic number of the specimen is constant in areas where the element of interest is not present. In other words, the method is upset if another heavy element does occur at a high concentration giving rise to a change in intensity of white x-ray radiation. This latter factor seems to narrow the potentials of HALL's method.

7.5.2.4. Absolute Standardization

The relative amounts or relative weight fractions of an element in the histological and cytological regions of a specimen is the kind of information that is often sought. Conversion of relative quantitative data to absolute units will add considerably to the potential value of the data obtained. For absolute standardization, reference standards are needed which are in chemical composition and physical properties as similar to the specimen as possible. With regard to biological specimens no general ways have been found thus far to achieve this aim. ANDERSEN (1967) has specified that a suitable standard must be well known chemically, homogeneous on the micron level, and able to withstand the high vacuum and heat loading imposed by EPA. He recommends to use organic compounds of high melting point containing a known concentration of a particular element, or pressed powder pellets of freeze-dried solutions in which the element of interest is dissolved in an aqueous solution of an organic matrix that contains the major elements found in the specimen in approximately the same concentration. Gelatin may represent such an organic matrix.

Both types of reference standards have been employed in EPA of animal specimens. Mixtures of the epoxy resins Epon 812 and 826 having known concentrations of Cl were used for absolute determinations of Cl (INGRAM and HOGBEN, 1968). ROBISON and DAVIS (1969) applied pellets containing organically bound I for analysis of I in thyroid glands, and ROSENSTIEL et al. (1970) used thin sections of either NaCl embedded in methacrylate or agar-agar containing known concentrations of Na, K and Cl as reference standards in electrolyte studies on kidney.

In a study on the accumulation of K in guard cells of tobacco leaves, SAWHNEY and ZELITCH (1969) used membrane filters as standards which had been impregnated with known concentrations of KCl and then freeze-dried. These filters had a thickness of 89 μ. Assuming that their density was about the same as that of freeze-dried plant material, i.e., 0.6 g/ml, the depth of electron penetration was approximately 16 μ in the standards and the epidermal preparations at 20 KeV accelerating potential (cf. Fig. 7.3). The thickness of such epidermal preparations is not uniform, however, and is likely to be less than 16 μ in certain regions. Thus, the volumes of specimen and reference standard analyzed were not always identical, and an error in standardization may have been made in those analyses. In a similar investigation HUMBLE and RASCHKE (1971) sought to overcome these difficulties through calibrating the relative data of K analyses with C coated microcrystals of K-oxalate and KCl, the volumes of which had been determined microscopically. Since in their electron probe analyses the beam penetrated both the specimens and the standards completely, the error in standardization appears eliminated.

Absolute standardization in EPA of plant specimens obviously requires a separate experimental study for each different problem, and no general recommendation can be made as to any particular material that may be used as reference standard.

7.5.3. Correlation of Elemental Distribution with Structures

Biological specimens do not show much contrast when they are viewed with the reflected light microscope of the electron probe analyzer. Accordingly, it is good practice to subject the specimen to a careful optical examination prior to analysis. It is often necessary to supplement the structural evaluation of the specimen with transmitted light microscopy of monitor sections.

The display of absorbed or backscattered electron images greatly helps in recognizing distinct structures and subsequently facilitates the correlation of structures with data from point analyses.

In line scanning, a relative measure of the concentration of an element is recorded as a function of distance. The recorded line scan must be correlated with the scanning line which is visible on the specimen as a result of exposure to the electron beam (see LÄUCHLI and SCHWANDER, 1966, Fig. 2a, and LÄUCHLI, SPURR and WITTKOPP, 1970, Fig. 1 for scanning lines photographed by reflected light microscopy). The following methods are available for correlating line scans with biological structures. If the specimen is moved under the static beam and the line scan recorded with a strip chart recorder, adequate structural correlation may be obtained by following the position of the beam in the tissue and simultaneously marking it on the line scan. A precise correlation is achieved after the line scan has been recorded, if the course of the scanning line on the specimen is evaluated with a reflected light microscope in polarized light. The scanning line is then clearly visible as a dark line. Correlation of the course of the line with the recorded concentration profile is made possible by quantitative measurements with an ocular micrometer (LÄUCHLI, 1967c).

If the electron beam is driven across the stationary specimen and the line scan displayed on an oscilloscope, a permanent record of the scanning line is obtained by photographing the electron image in the absorbed or backscattered mode from another oscilloscope at the same magnification. This method permits a very accurate correlation with the concentration profile that appears on the first oscilloscope (cf. WAISEL, HOFFEN and ESHEL, 1970, Fig. 1).

7.6. *Applications to Plant Physiology*

7.6.1. Mineral Nutrition: Physiological and Ecological Aspects

An evaluation of the literature suggests that electron probe analysis (EPA) has had its greatest impact on research in mineral nutrition of plants. It has been shown that the lateral transport of ions into the xylem of roots can be investigated effectively by EPA (LÄUCHLI, 1967c, 1968a; LÄUCHLI, SPURR and EPSTEIN, 1971). Translocation of inorganic solutes in xylem and phloem as well as their lateral transfer between the two principal conducting tissues has also been studied with this method (LÄUCHLI and SCHWANDER, 1966; LÄUCHLI, 1967c, 1968b). Furthermore, the pattern of distribution of mineral elements in any plant tissue may be established using the electron probe analyzer.

Specific tracing of solute transport in xylem and phloem is achieved through analyses of the sap of xylem vessels and sieve tubes. A method has been described for collecting xylem exudate from roots which permits precise kinetic studies of ion transport into the xylem of roots (LÄUCHLI and EPSTEIN, 1971). This method may possibly be adapted for collection of xylem-vessel sap from stems, petioles and fruit stalks. On the other hand, the aphid-stylet technique allows the collection of uncontaminated sieve-tube sap (KENNEDY and MITTLER, 1953). Both xylem and phloem sap could be subjected to a complete elemental analysis with the electron probe analyzer (cf. INGRAM and HOGBEN, 1967; LECHENE, 1970).

High concentrations of mineral elements in cells, be they as solutes or as mineral deposits, are to be determined most accurately by means of EPA. This method has contributed to the discovery of the physiological function of K in cells which control movements through turgor changes, that is stomatal movement (SAWHNEY and ZELITCH, 1969; HUMBLE and RASCHKE, 1971) and nyctinasty (SATTER, MARINOFF and GALSTON, 1970). Further, the benefit of EPA for the detection of mineral deposits has been demonstrated by several investigators (LÄUCHLI, 1967b, c, 1968b; LAROCHE, 1967; RASMUSSEN, 1968; ROLAND and BESSOLES, 1968; KAUFMANN et al., 1969). Microcrystals and amorphous deposits of minerals in plants appear to be characterized chemically by the electron probe analyzer with more efficiency than by crystallographic methods. In addition, it is possible to determine both the cationic and anionic components of a mineral deposit as has been shown for lichens containing large amounts of Fe-compounds (NOESKE et al., 1970).

Not only has EPA proved itself to be applicable to the physiological aspects of mineral nutrition, but it may also have a great impact upon ecological research. There are innumerable ecological problems in mineral nutrition of plants (see EPSTEIN, 1972). Soil, the substrate from which land plants acquire their essential mineral nutrients, often contains substances that adversely affect plant growth. Vast areas of the earth are unsuitable for growth of many plant species including most of the agricultural crops because their soils are saline, the concentration of salt (NaCl) being high. Only the halophytes are capable of tolerating high salt concentrations which impair growth of other plants. Mineral nutrition of halophytes on the one hand and salt tolerance of non-halophytic species on the other are problems of broad interest. Electron probe analysis offers an intriguing method to inquire into certain aspects of salinity such as the area of absorption, translocation and secretion of Na.

Some soils are affected by high concentrations of heavy metals, thus excluding certain plant species or causing metal toxicity in others. A better understanding of the physiological basis of metal toxicity in plants may be provided through measuring the respective metals in the tissue with the electron probe analyzer. Such an approach has been chosen by RASMUSSEN (1968) and by WAISEL, HOFFEN and ESHEL (1970) to study Al toxicity in higher plants. It could even be expanded to environmental pollutants like Pb (cf. CARROLL, SPINELLI and GOYER, 1970) and Hg which did not normally occur in plants to any great amounts until recently when the concentration of these poisonous metals in the biosphere was found to rise in an alarming manner.

7.6.2. Cellular Compartmentation

Cryostat sections which have been used in the majority of the investigations of plant materials do not permit one to clearly distinguish between the various subcellular compartments except in special cases like the leaf cells of *Mnium* where K and P could be detected separately in two compartments, the cytoplasm (including cytoplasmic organelles) and the vacuole (LÄUCHLI and LÜTTGE, 1968). However, the use of thin sections will lead to an improved spatial resolution in EPA, thus allowing compartmentation studies.

With regard to ion absorption, much work on compartmentation of plant cells has been conducted with the giant cells of certain algae (cf. MACROBBIE, 1971). The methods used in that work allowed, at best, a distinction between cell wall, cytoplasmic layer and vacuole. Electron probe analysis would appear to extend compartmentation in these giant cells to include the cytoplasmic organelles nucleus, chloroplasts and probably mitochondria. Further, similar compartmentation studies may be made in the much smaller cells of higher plants because it appears possible to improve the resolution in EPA to about 0.1 μ (HALL, 1968b).

In single cells of bean stem tissue in shake culture, BAJAJ, RASMUSSEN and ADAMS (1971) have carried out a compartmentation study indicating that Ca is distributed relatively uniformly while the elements K, P and Zn appeared accumulated in the nucleus. LIBANATI and TANDLER (1969) determined the intracellular distribution of phosphate ions *in situ* using 1 μ sections of corn root tips after having precipitated the soluble phosphate with Pb-acetate. Regardless of whether or not a complete retention of phosphate ions was achieved in that study (cf. discussion of precipitation technique in section on Preparation Techniques), Fig. 7.9 does show that P is accumulated within the nucleolus of the meristematic root cells. Fig. 7.9a represents an absorbed electron image featuring a distinct dark phase that corresponds with a nucleolus. The dark phase, indicative of a high average atomic number, is caused by the elements P (Fig. 7.9b) and Pb (Fig. 7.9c). Sulfur (Fig. 7.9d) is not concentrated in the nucleolus in a "control" section of a root segment that was fixed in glutaraldehyde only.

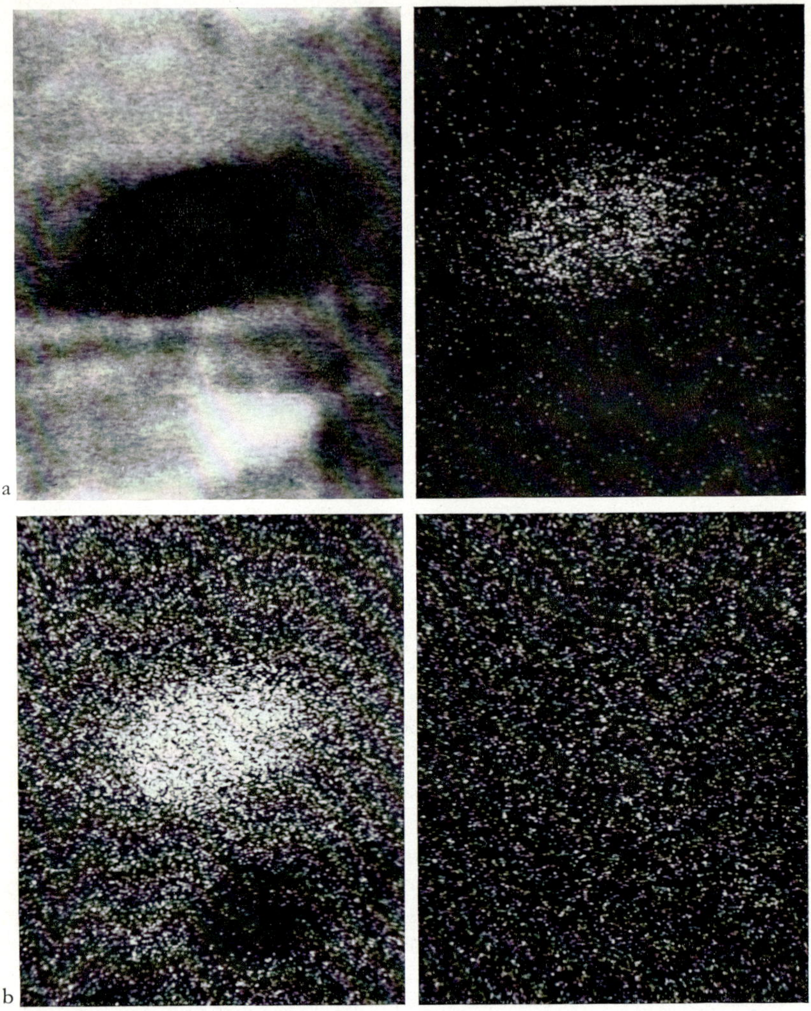

Fig. 7.9. Two-dimensional scan from a 1 μ section of a corn root tip depicting the nuclear region of a meristematic cell. Root segment fixed in Pb-acetate, pH 6.0, dehydrated and embedded in Maraglas. Section mounted on a Cu grid and coated on both sides with 100 Å Al. a: absorbed electron image. b to d: x-ray images of P(K_α), Pb(L_α), S(K_α) radiation respectively. d is taken from a "control" section of a root segment fixed in gutaraldehyde. × 3,800. (From LIBANATI and TANDLER, 1969)

A different approach for yielding information on compartmentation of plant cells also appears feasible, that is EPA of isolated cell organelles. Isolation of such organelles as chloroplasts and mitochondria is performed routinely in many laboratories. It offers an alternative method especially if non-aqueous isolation media are used.

7.6.3. Microautoradiography and Electron Probe Analysis

A comparative literature search by LÄUCHLI (1968a) has revealed that results concerning the lateral transport of ions into the xylem of roots obtained through EPA can be reconciled with those from microautoradiographic studies. Yet, there is a basic difference in the analytical data that are gained with the two methods. Electron probe analysis provides information on the relative amounts of a chemical element per volume analyzed; microautoradiography, however, yields data on the relative amounts of a radioactively labeled element per specimen volume. In other words, the two methods produce data on chemical amounts and activities respectively. An experimental combination of both methods would consequently lead to data representing specific activities. Thus far, the measurement of specific activities in microscopically small tissue regions has not been accomplished, but it does offer attractive new possibilities for plant physiological research.

A preliminary attempt to correlate microautoradiography and EPA was made when LÄUCHLI and LÜTTGE (1968) examined the kinetics of ion absorption into the cytoplasmic layer of *Mnium* leaf cells as a function of ion concentration in the medium. The absorption of $^{35}SO_4^{--}$ was followed by microautoradiographic detection of ^{35}S whereas EPA was used to pursue the absorption of K^+ into the cytoplasm. Activities of ^{35}S were measured after exposure of leaves for 3 hr to solutions containing $K_2{}^{35}SO_4$ at concentrations varying from 0.001 mM to 0.2 mM. Amounts of K were determined after an absorption period of 18 hr in solutions containing 0.05 mM to 0.5 mM KH_2PO_4. "Control" leaves were exposed for 18 hr to a solution containing only 0.1 mM $CaSO_4$. Net K^+ absorption was then calculated by subtracting the average amount of K in the cytoplasm of the control leaves from the respective values of the leaves exposed to KH_2PO_4.

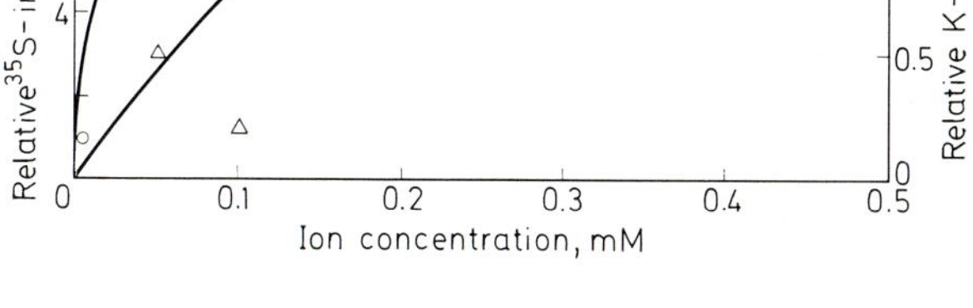

Fig. 7.10. Microautoradiographic detection of ^{35}S-activities and electron probe analysis of amounts of K in cytoplasm of leaf cells of *Mnium* as a function of the concentration of $K_2{}^{35}SO_4$ and KH_2PO_4 respectively (further details in text). (Redrawn from LÄUCHLI and LÜTTGE, 1968, Fig. 5 and 7)

The results of this comparative experiment are presented in Fig. 7.10 and reflect a relative measure of the kinetics of transport of $^{35}SO_4^{--}$ and K^+ across the plasmalemma into the cytoplasm. Fig. 7.10 indicates that the kinetics of transport of both ions as a function of concentration up to 0.5 mM may be described by a saturation curve. Such a relation has been found for many plant materials and for various ions when absorption by excised plant tissues was measured (EPSTEIN, 1966).

It must be emphasized that no specific activities were measured in the experiment shown in Fig. 7.10. The means were set out, however, by which this goal can be achieved.

7.6.4. Histochemical and Cytochemical Studies

Electron probe analysis not only permits the detection of chemical elements already present in the specimen but also appears useful in elucidating histochemical and cytochemical reactions for the demonstration of specific substances. The histochemical study of an enzymatic reaction shall serve as an example. When applied to an enzymatic reaction, EPA enables one to localize the primary, secondary and final reaction products of the histochemical "staining" method. It is superior to the light microscopic analysis of an enzymatic reaction because the invisible reaction products may be spotted chemically in the electron probe analyzer. In animal tissues the enzymes alkaline phosphatase (HALE, 1962) and EDTA-activated ATPase (ENGEL, RESNICK and MARTIN, 1968) have been localized in such a manner.

Nucleic acids are detectable by EPA. SIMS and MARSHALL (1966) have stained nucleic acids by means of a gallocyanin chrome alum chelate. The intracellular location of the nucleic acids was then examined by determining the distribution of Cr.

Histochemical procedures are in use to demonstrate the existence of certain structural components in biological specimens. Ectodesmata in the outer cell walls of leaves, for instance, are stained specifically by Gilson solution the essential component of which being $HgCl_2$ (LAMBERTZ, 1954). Electron probe analysis now has confirmed that the crystalline precipitates which are formed in the ectodesmata upon treatment with Gilson solution do contain the elements Hg and Cl (SCHÖNHERR and BUCOVAC, 1970).

7.7. *Related Methods*

7.7.1. Analytical Electron Microscopy

The present article has dealt thus far with the classical type of electron probe analyzer with which a tremendous amount of new information was gained, predominantly in mineralogy and metallurgy. The fields of application of this instrument broadened in the 1960's and do now encompass biology to a rapidly growing extent. The unique and complex properties of biological specimens on the one side and the desire of biologists to perform electron probe analysis (EPA) at ultramicroscopic dimensions on the other spurred the development of instruments whose specifications would be adapted to the biologists' needs.

There is now a trend to equip scanning electron microscopes with x-ray detectors. Tousimis (1969b, c) used such an instrument with dispersive detectors for work on animal material. He achieved spatial resolutions of 200 to 800 Å with the electron mode and between 0.1 and 1 μ for x-ray detection. Even better resolutions were obtained by Ogilvie and Sutfin (1970) with a scanning electron microscope equipped with a transmitted electron detector and a nondispersive x-ray detector. Working with 0.2 μ thick sections of animal material, these investigators were able to detect granules in mitochondria which showed a high P content and were only 0.05 μ in diameter. A resolution of 0.05 μ puts EPA at the upper limit of ultramicroscopic dimensions.

Analytical electron microscopy evolved directly from the concept of the classical electron probe analyzer. Basically, an analytical electron microscope is a transmission electron microscope in combination with one or several x-ray detectors. Galle in France used an analytical electron microscope for biological work with a low magnification power up to $\times$ 15,000 (Galle, 1966; Galle and Stuve, 1968). AEI Scientific Apparatus Ltd., Harlow, Essex, England designed the analytical electron microscope EMMA 4 which may be the preferred instrument for biological specimens. It is a high-resolution transmission electron microscope featuring two dispersive and one nondispersive detector for x-ray analysis. The operation of the analytical electron microscope and its possible applications were first described by Duncumb (1966). A description of EMMA 4, which is now commercially available, was given by Cooke and Openshaw (1970) and makes clear the outstanding capabilities of this instrument. The point to point microscope resolution of EMMA 4 is 10 Å, x-ray analysis may be conducted from regions 0.1 μ in diameter, and — last but not least — the detection sensitivity is claimed to be 10^{-17} g, a considerable improvement over the sensitivity of the classical electron probe analyzer (cf. Table 7.1).

That analytical electron microscopy provides for accurately determining the location and concentration of chemical elements in subcellular regions of plant specimens has been demonstrated by Chandler and Chou (1970). It was shown that Cu-Cr-arsenate, a wood preservative effective against fungal attack, is accumulated in the cell walls of wood with the x-ray intensity peaks of the three elements corresponding to the region of the middle lamella. Surely, one has to agree with the author's conclusion that "this experiment demonstrates the unique ability of the technique to perform subcellular elemental analyses which will prove invaluable in many other biological studies" (cf. Hodson and Marshall, 1970b; Höhling et al., 1970; Podolsky, Hall and Hatchett, 1970; Schafer and Chandler, 1970).

7.7.2. Laser Probe Analysis

In 1963 Rosan, Healy and McNary have reported on spectroscopic ultramicroanalysis with a laser, a new method which in some degree is related to EPA. In this method a focused laser is discharged onto a specimen, forming a crater from which energized gases are produced. The gas cloud being characteristic of the elements that have been vaporized is made to set off a spark discharge between charged electrodes. Light from this spark discharge passes through a spectrograph for photographic recording of the lines of the atomic spectra thereby permitting a complete qualitative and semiquantitative analysis of the vaporized fraction of the specimen (see Glick, 1966, for a complete description). A laser probe analyzer is manufactured by Jarrel-Ash Company, Waltham, Massachusetts 02154.

The sensitivity of this method was determined to be 10^{-11} g for Mg (Rosan, Brech and Glick, 1965). Spatial resolution was improved in the last years so that analyses may be performed at cellular dimensions (Peppers et al., 1968).

Laser probe analysis shows one particular feature which gives this method unique potentials to biological applications, that is the possibility of *in vivo* analysis. Whereas completely dehydrated specimens are required for EPA, the laser probe analyzer permits a chemical analysis in microscopic areas of surfaces of living tissues, organs and organisms.

7.7.3. Ion Probe Mass Analysis

The analytical capabilities of EPA, analytical electron microscopy and laser probe analysis are by far surpassed by ion probe mass analysis — according to Bayard (1969a) a method which sounds almost like science fiction. This exciting new analytical tool goes back to the work of Castaing and SLODZIAN (1962). To build the ion probe mass analyzer Castaing took over the basic concepts of the electron probe analyzer but used an ion beam instead of an electron beam and replaced the crystal spectrometers by a mass spectrometer.

Applied Research Laboratories, Sunland, California 91040 has just begun to produce commercially the ion probe mass analyzer IMMA. In IMMA a primary ion beam consisting of ions of oxygen or an inert gas is focused on the specimen. Secondary ions from the sample are then formed, collected, and analyzed in a mass spectrometer. Accordingly, the method not only permits detection of all of the elements in the periodic table but also all of the isotopes. The ion beam in IMMA is continuously adjustable from 2 μ to 300 μ in diameter. The nature of the ion probe mass analyzer renders the method destructive whereas EPA is virtually nondestructive even when applied to biological samples.

Ion probe mass analysis shows a detection sensitivity of 10^{-20} g (see Table 7.1) surpassing that of EPA by at least four orders of magnitude. GALLE, BLAISE and SLODZIAN (1969) indeed were able to detect as little as 10^{-20} g Na in a smear of human red blood cells, corresponding to about 300 Na atoms!. According to BAYARD (1969a), the investigator who is fortunate to have access to the ion probe mass analyzer will be able to detect from 100 to 1000 atoms. In other terms, he may analyze successive ten-atom-thick sections of his specimen!

7.8. *Conclusions*

Until 1960, electron probe analysis (EPA) was applied almost exclusively to inanimate materials. The fields of application were widened in the 1960's to include biological specimens. The number of publications in which EPA of biological materials was described amounted to 82 by the end of 1968 (HALL and HÖHLING, 1969). Since that time, the new method experienced a break-through into many areas of the biological sciences including plant physiology. This took place despite the fact that investigators had to use instruments which were not specifically designed for biological samples and were extremely expensive. It is to be expected that EPA will be accepted among the important analytical tools in plant physiology. One may speculate further that instruments like the analytical electron microscope and the scanning electron microscope with transmitted mode and x-ray detectors will hasten that development because of extension of the method's capabilities into the ultramicroscopic dimension.

If the present chapter helps to stimulate the interest of plant scientists who had been unaware of EPA, its aim will be achieved.

References

Aei Scientific Apparatus Inc.: EMMA4 — electron microscopy/x-ray microanalysis. Science **169** A, 155 A (1970).

Albright, J., Lewis, R.: Preparation of biological specimens for electron-beam microanalyzer. J. appl. Phys. **36**, 2615 (1965).

Andersen, C. A.: An introduction to the electron probe microanalyzer and its application to biochemistry. Meth. biochem. Anal. **15**, 147—270 (1967).

Andersen, C. A., Hasler, M. F.: Extension of electron microprobe techniques to biochemistry by the use of long wavelength x-rays. In: X-ray optics and microanalysis, R. Castaing, P. Deschamps and J. Philibert, eds., p. 310—327. Paris: Herrmann 1966.

Appleton, T. C.: Ultra-thin frozen sections for electron microscopy. LKB Instruments, Inc., Preliminary Report (1968).

Bajaj, Y. P., Rasmussen, H. P., Adams, M. W.: Electron microprobe analysis of isolated plant cells. J. Exp. Bot. **22**, 749—752 (1971).

Bayard, M.: Microprobes. Amer. Lab. 1, 8—12 (1969a).

Bayard, M.: Computer interfacing the electron microprobe. Microsc. **17**, 169—177 (1969b).

Biddulph, S. F.: Visual indications of S^{35} and P^{32} translocation in the phloem. Amer. J. Bot. **43**, 143—148 (1956).

Birks, L. S.: Electron probe microanalysis. New York-London: Interscience Publ. 1963.

Böck, P.: Elektronenmikroskopischer Nachweis von Na^{+}, Ca^{++} und Cl^{-} in der lactierenden Milchdrüse des Meerschweinchens. Cytobiologie **2**, 68—82 (1970).

Branton, D., Jacobson, L.: Freeze-drying of plant material. Exp. Cell Res. **22**, 559—568 (1961).

Campbell, A. J., Gibbons, R.: Specimen contamination in the electron microprobe. In: The electron microprobe, T. D. McKinley, K. F. J. Heinrich and D. B. Wittry, eds., p. 75—82. New York: Wiley and Sons 1966.

Carroll, K. G., Spinelli, F. R., Goyer, R. A.: Electron probe microanalyzer localization of lead in kidney tissue of poisoned rats. Nature **227**, 1056 (1970).

Castaing, R.: Application des sondes électroniques à une méthode d'analyse ponctuelle chimique et cristallographique. Thèse science, Paris (1951).

Castaing, R.: Electron probe microanalysis. Adv. Electronics and Electron Phys. **13**, 317—386 (1960).

Castaing, R., Slodzian, G.: Microanalyse par émission ionique secondaire. J. Microsc. **1**, 395—410 (1962).

Cescas, M. P., Tyner, E. H., Gray, L. J.: The electron microprobe x-ray analyzer and its use in soil investigations. Adv. Agron. **20**, 153—198 (1968).

Chandler, J. A., Chou, C. K.: Subcellular elemental analysis with the analytical electron microscope. Proc. 28th ann. meet. electron microsc. soc. amer., Houston, 1970, C. J. Arceneaux, ed., p. 74—75. Baton Rouge: Claitor's Publ. Div. 1970.

Chase, G. D., Rabinowitch, J. L.: Principles of radioisotope methodology, 2nd ed. Minneapolis: Burgess 1962.

Chayen, J., Cunningham, G. J., Gahan, P. B., Silcox, A. A.: Life-like preservation of cytoplasmic detail in plant cells. Nature **186**, 1068—1069 (1960).

Clarkson, D. T., Sanderson, J.: A critical review of certain autoradiographic techniques. Agricult. Res. Council ARCRL. **20**, 68—73 (1970).

Constantin, L. L., Franzini-Armstrong, C., Podolsky, R. J.: Localization of calcium-accumulating structures in striated muscle fibers. Science **147**, 158—160 (1965).

Cooke, C. J., Openshaw, I. K.: Combined high resolution electron microscopy and x-ray microanalysis. Proc. 28th ann. meet. electron. microsc. soc. amer., Houston, 1970, C. J. Arceneaux, ed., p. 552—553. Baton Rouge: Claitor's Publ. Div. 1970.

Cosslett, V. E., Duncumb, P.: Microanalysis by a flying-spot x-ray method. Nature **177**, 1172—1173 (1956).

Crossett, R. N.: Autoradiography of ^{32}P in maize roots. Nature **213**, 312—313 (1967).

Davidson, E., Neuhaus, H.: Sample image displays for qualitative microprobe analysis. 1st natl. conf. on electron probe microanalysis, College Park, Maryland, 1966.

Duncumb, P.: Precipitation studies with EMMA — a combined electron microscope and x-ray microanalyzer. In: The electron microprobe, T. D. McKinley, K. F. J. Heinrich and D. B. Wittry, eds., p. 490—499. New York: Wiley and Sons 1966.

Engel, W. K., Resnick, J. S., Martin, E.: The electron probe in enzyme histochemistry. J. Histochem. Cytochem. **16**, 273—275 (1968).

Engström, A.: X-ray methods in histochemistry. Physiol. Rev. **33**, 190—201 (1953).

Epstein, E.: Dual pattern of ion absorption by plant cells and by plants. Nature **212**, 1324—1327 (1966).

Epstein, E.: Mineral nutrition of plants: principles and perspectives. New York: Wiley and Sons (1972).

Eschrich, W.: Translokation ^{14}C-markierter Assimilate im Licht und im Dunkeln bei *Vicia faba*. Planta **70**, 99—124 (1966).

Feder, N., O'Brien, T. P.: Plant microtechnique: some principles and new methods. Amer. J. Bot **55**, 123—142 (1968).

Gahan, P. B., Greenoak, G. C., James, D.: Preparation of ultra-thin frozen sections of plant tissues for electron microscopy. Histochemie **24**, 230—235 (1970).

Gahan, P. B., McLean, J., Kalina, M., Sharma, W.: Freezing-sectioning of plant tissues: the technique and its use in plant histochemistry. J. Exp. Bot. **18**, 151—159 (1967).

Galle, P.: Cytochimie sur coupes ultrafines par spectrographie des rayons x. In: 6th int. congr. electron microscopy, Kyoto, 1966, p. 79—80. Tokyo: Maruzen 1966.

Galle, P., Blaise, G., Slodzian, G.: Study of distribution of diffusible ions in biological tissues using a secondary ion emission microanalyzer. Proc. 4th natl. conf. electron microprobe analysis, Pasadena, 1969, p. 36. Electron Probe Analysis Society of America, 1969.

Galle, P., Stuve, J.: Analyse par microsonde électronique sur coupes biologiques ultrafines. J. Microsc. **7**, 590—592 (1968).

Gielink, A. J., Sauer, G., Ringoet, A.: Histoautoradiographic localization of calcium in oat plant tissues. Stain Technol. **41**, 281—286 (1966).

Gleit, C. E., Holland, W. D.: Use of electrically excited oxygen for the low temperature decomposition of organic substances. Anal. Chem. **34**, 1454—1457 (1962).

Glick, D.: The laser microprobe: its use for elemental analysis in histochemistry. J. Histochem. Cytochem. **14**, 862—868 (1966).

Hale, A. J.: Identification of cytochemical reaction products by scanning x-ray emission microanalysis. J. Cell Biol. **15**, 427—435 (1962).

Hall, T.: Some aspects of the microprobe analysis of biological specimens. In: Quantitative electron probe microanalysis, K. F. J. Heinrich, ed., Spec. Publ. No. **298**, 269—299. Washington: Natl. Bureau of Standards 1968a.

Hall, T.: Electron probe x-ray microanalysis. Proc. Roy. Microsc. Soc. **3**, 84—85 (1968b).

Hall, T. A., Hale, A. J., Switsur, V. R.: Some applications of microprobe analysis in biology and medicine. In: The electron microprobe, T. D. McKinley, K. F. J. Heinrich and D. B. Wittry, eds., p. 805—833. New York: Wiley and Sons 1966.

HALL, T. A., HÖHLING, H. J.: The application of microprobe analysis to biology. In: 5th int. congr. x-ray optics microanalysis, Tübingen, 1968, G. MÖLLENSTEDT and K. H. GAUKLER, eds., p. 581—591. Berlin-Heidelberg-New York: Springer 1969.

HALL, T. A., WERBA, P.: The measurement of total mass per unit area and elemental weight-fractions along line scans in thin specimens. In: 5th int. congr. x-ray optics microanalysis, Tübingen, 1968, G. MÖLLENSTEDT and K. H. GAUKLER, eds., p. 93—98. Berlin-Heidelberg-New York: Springer 1969.

HEINRICH, K. F. J.: Electron probe microanalysis: a review. Appl. Spectrosc. **22**, 395—403 (1968).

HODSON, S., MARSHALL, J.: Ultramicrotomy: a technique for cutting ultrathin sections of unfixed frozen biological tissues for electron microscopy. J. Microsc. **91**, 105—117 (1970a).

HODSON, S., MARSHALL, J.: Tissue sodium and potassium: direct detection in the electron microscope. Experientia **26**, 1283—1284 (1970b).

HÖHLING, H. J., HALL, T. A.: Elektronenstrahlmikroanalyse als quantitative histologische Methode. Naturwissenschaften **56**, 622—629 (1969).

HÖHLING, H. J., SCHÖPFER, H., HÖHLING, R. A., HALL, T. A., GIESEKING, R.: The organic matrix of developing tibia and femur, and macromolecular deliberations. Naturwissenschaften **57**, 357 (1970).

HUMBLE, G. D., RASCHKE, K.: Stomatal opening quantitatively related to potassium transport. Plant Physiol. **48**, 447—453 (1971).

INGRAM, F. D.: A small computerized on-line data gathering system for the electron microprobe. Proc. 4th natl. conf. electron microprobe analysis, Pasadena, 1969, p. 27. Electron Probe Analysis Society of America, 1969.

INGRAM, M. J., HOGBEN, C. A. M.: Electrolyte analysis of biological fluids with the electron microprobe. Anal. Biochem **18**, 54—57 (1967).

INGRAM, M. J., HOGBEN, C. A. M.: Procedures for the study of biological soft tissue with the electron microprobe. Develop. Appl. Spectrosc. **6**, 43—54 (1968).

JENSEN, W. A.: Botanical histochemistry. Principles and practice. San Francisco and London: W. H. Freeman 1962.

KAUFMANN, P. B., BIGELOW, W. C., PETERING, L. B., DROGOSZ, F. B.: Silica in developing epidermal cells of *Avena* internodes: electron microprobe analysis. Science **166**, 1015—1017 (1969).

KEIL, K.: The electron microprobe x-ray analyzer and its application in mineralogy. Fortschr. Miner. **44**, 4—66 (1967).

KEMP, J. W.: X-ray wavelength tables. Glendale, Calif.: Appl. Res. Lab. 1962.

KENNEDY, J. S., MITTLER, T. E.: A method of obtaining phloem sap via the mouth parts of aphids. Nature **171**, 528 (1953).

KNOX, R. B.: Freeze-sectioning of plant tissues. Stain Technol. **45**, 265—272 (1970).

KOMNICK, H.: Elektronenmikroskopische Lokalisation von Na^+ und Cl^- in Zellen und Geweben. Protoplasma **55**, 414—418 (1962).

KOMNICK, H., KOMNICK, U.: Elektronenmikroskopische Untersuchungen zur funktionellen Morphologie des Ionentransportes in der Salzdrüse von *Larus argentatus*, V. Experimenteller Nachweis der Transportwege. Z. Zellforsch. **60**, 163—203 (1963).

KRICHBAUM, R., LÜTTGE, U., WEIGL, J.: Mikroautoradiographische Untersuchung der Auswaschung des „anscheinend freien Raumes" von Maiswurzeln. Ber. dtsch. bot. Ges. **80**, 167—176 (1967).

LAMBERTZ, P.: Untersuchungen über das Vorkommen von Plasmodesmen in den Epidermisaußenwänden. Planta **44**, 147—190 (1954).

LANE, B. P., MARTIN, E.: Electron probe analysis of cationic species in pyroantimonate precipitates in epon-embedded tissue. J. Histochem. Cytochem. **17**, 102—106 (1969).

LAROCHE, J.: Localisation de la silice par le microanalyseur à sonde électronique. C. R. Acad. Sc. (Paris) D **265**, 1695—1697 (1967).

LÄUCHLI, A.: Cryostat technique for fresh plant tissues and its application in enzyme histochemistry. Planta **70**, 13—25 (1966).

LÄUCHLI, A.: Zur Technik der Herstellung biologischer Präparate für die histochemische Untersuchung mit der Röntgen-Mikrosonde. Histochemie **11**, 286—288 (1967a).

LÄUCHLI, A.: Nachweis von Calcium-Strontium-Ablagerungen im Fruchtstiel von *Pisum sativum* mit der Röntgen-Mikrosonde. Planta **73**, 221—227 (1967b).

LÄUCHLI, A.: Untersuchungen über Verteilung und Transport von Ionen in Pflanzengeweben mit der Röntgen-Mikrosonde. I. Versuche an vegetativen Organen von *Zea mays*. Planta **75**, 185—206 (1967c).

LÄUCHLI, A.: Untersuchung des Stofftransports in der Pflanze mit der Röntgen-Mikrosonde. Vortr. Gesamtgeb. Bot. N. F. **2**, 58—65 (1968a).

LÄUCHLI, A.: Untersuchungen mit der Röntgen-Mikrosonde über Verteilung und Transport von Ionen in Pflanzengeweben. II. Ionentransport nach den Früchten von *Pisum sativum*. Planta **83**, 137—149 (1968b).

LÄUCHLI, A., EPSTEIN, E.: Lateral transport of ions into the xylem of corn roots. I. Kinetics and energetics. Plant Physiol. **48**, 111—117 (1971).

LÄUCHLI, A., LÜTTGE, U.: Untersuchung der Kinetik der Ionenaufnahme in das Cytoplasma von *Mnium*-Blattzellen mit Hilfe der Mikroautoradiographie und der Röntgen-Mikrosonde. Planta **83**, 80—98 (1968).

LÄUCHLI, A., SCHWANDER, H.: X-ray microanalyzer study on the localization of minerals in native plant tissue sections. Experientia **22**, 503—505 (1966).

LÄUCHLI, A., SPURR, A. R., EPSTEIN, E.: Lateral transport of ions into the xylem of corn roots. II. Evaluation of a stelar pump. Plant Physiol. **48**, 118—124 (1971).

LÄUCHLI, A., SPURR, A. R., WITTKOPP, R. W.: Electron probe analysis of freeze-substituted, epoxy resin embedded tissue for ion transport studies in plants. Planta **95**, 341—350 (1970).

LECHENE, C.: The use of the electron microprobe to analyse very minute amounts of liquid samples. Proc. 5th natl. conf. electron probe analysis, New York City, 1970, p. 32A—32C. Electron Probe Analysis Society of America, 1970.

LENNEP, E. W. VAN, KOMNICK, H.: Histochemical demonstration of sodium and chloride in the frog epidermis. Cytobiologie **3**, 137—151 (1971).

LIBANATI, C. M., TANDLER, C. J.: The distribution of the water-soluble inorganic orthophosphate ions within the cell: accumulation in the nucleus. J. Cell Biol. **42**, 754—765 (1969).

LÜTTGE, U., WEIGL, J.: Mikroautoradiographische Untersuchungen der Aufnahme und des Transports von $^{35}SO_4^{--}$ und $^{45}Ca^{++}$ in Keimwurzeln von *Zea mays* L. und *Pisum sativum* L. Planta **58**, 113—126 (1962).

LÜTTGE, U., WEIGL, J.: Zur Mikroautoradiographie wasserlöslicher Substanzen. Planta **64**, 28—36 (1965).

MACROBBIE, E. A. C.: Fluxes and compartmentation in plant cells. Ann. Rev. Plant Physiol. **22**, 75—96 (1971).

MARTON, L.: Survey of microanalysis — interpolation and extrapolation. In: Electron probe microanalysis, A. J. TOUSIMIS and L. MARTON, eds., p. 1—14. New York: Academic Press 1969.

MCKINLEY, T. D., HEINRICH, K. J. F., WITTRY, D. B., eds.: The electron microprobe. New York: Wiley and Sons 1966.

NOESKE, O., LÄUCHLI, A., LANGE, O. L., VIEWEG, G. H., ZIEGLER, H.: Konzentration und Lokalisierung von Schwermetallen in Flechten der Erzschlackenhalden des Harzes. Dtsch. Bot. Ges. N. F., Nr. **4**, 67—79 (1970).

Ogilvie, R. E., Sutfin, L. V.: New techniques in scanning electron microscopy. Proc. 28th ann. meet. electron microsc. soc. amer., Houston, 1970, C. J. Arceneaux, ed., p. 348—349. Baton Rouge: Claitor's Publ. Div. 1970.

Ong, P. S.: Analysis for low atomic number elements with the electron microprobe. In: Electron probe microanalysis, A. J. Tousimis and L. Marton, eds., p. 137—153. New York: Academic Press 1969.

Osmond, C. B., Lüttge, U., West, K. R., Pallaghy, C. K., Shacher-Hill, B.: Ion absorption in *Atriplex* leaf tissue. II. Secretion of ions to epidermal bladders. Aust. J. Biol. Sci. **22**, 797—814 (1969).

Pease, D. C.: Anhydrous ultrathin sectioning and staining for electron microscopy. J. Ultrastruct. Res. **14**, 379—390 (1966).

Peppers, N. A., Scribner, E. J., Alterton, L. E., Honey, R. C., Beatrice, E. S., Harding-Barlow, I., Rosan, R. C., Glick, D.: Q-switched ruby laser for emission microspectroscopic elemental analysis. Anal. Chem. **40**, 1178—1182 (1968).

Pickering, E. R.: Autoradiography of mobile, C^{14}-labeled herbicides in sections of leaf tissue. Stain Technol. **41**, 131—137 (1966).

Podolsky, R. J., Hall, T., Hatchett, S. L.: Identification of oxalate precipitates in striated muscle fibers. J. Cell Biol. **44**, 699—702 (1970).

Rains, D. W., Läuchli, A.: Unpublished data (1969).

Rasmussen, H. P.: Entry and distribution of aluminum in *Zea mays*. Planta **81**, 28—37 (1968).

Rasmussen, H. P.: Unpublished data (1969).

Rasmussen, H. P., Shull, V. E., Dryer, H. T.: Determination of element localization in plant tissue with the microprobe. Develop. Appl. Spectrosc. **6**, 29—42 (1968).

Robison, W. L., Davis, D.: Determination of iodine concentration and distribution in rat thyroid follicles by electron-probe microanalysis. J. Cell Biol. **43**, 115—122 (1969).

Roland, J. C., Bessoles, M.: Mise en évidence du calcium dans les cellules du collenchyme. C. R. Acad. Sc. (Paris) D **267**, 589—592 (1968).

Rosan, R. C., Brech, F., Glick, D.: Current problems in laser microprobe analysis. Fed. Proc. **24**, S126—S128 (1965).

Rosan, R. C., Healy, M. K., McNary, W. F., Jr.: Spectroscopic ultramicroanalysis with a laser. Science **142**, 236—237 (1963).

Rosenstiel, A. V. P., Höhling, H. J., Schermann, J., Kriz, W.: Electron probe microanalysis of electrolytes in kidney slices. Proc. 5th natl. conf. electron probe analysis, New York City, 1970, p. 33A—33B. Electron Probe Analysis Society of America, 1970.

Sanui, H., Pace, N.: Application of atomic absorption in the study of Na, K, Mg and Ca binding by cellular membranes. Appl. Spectrosc. **20**, 135—141 (1966).

Satter, R. L., Marinoff, P., Galston, A. W.: Phytochrome controlled nyctinasty in *Albizzia julibrissin*. II. Potassium flux as a basis for leaflet movement. Amer. J. Bot. **57**, 916—926 (1970).

Sawhney, B. L., Zelitch, I.: Direct determination of potassium ion accumulation in guard cells in relation to stomatal opening in light. Plant Physiol. **44**, 1350—1354 (1969).

Schafer, P. W., Chandler, J. A.: Electron probe x-ray microanalysis of a normal centriole. Science **170**, 1204—1205 (1970).

Schönherr, J., Bukovac, M. J.: The nature of precipitates formed in the outer cell wall following fixation of leaf tissue with Gilson solution. Planta **92**, 202—207 (1970).

Shone, M. G. T., Clarkson, D. T., Sanderson, J.: The absorption and translocation of sodium by maize seedlings. Planta **86**, 301—314 (1969).

Sims, R. T., Hall, T.: X-ray emission microanalysis of proteins and sulphur in rat plantar epidermis. J. Cell Sci. **3**, 563—572 (1968).

SIMS, R. T., MARSHALL, D. J.: Location of nucleic acids by electron probe x-ray microanalysis. Nature **212**, 1359 (1966).

SPURR, A. R.: A low-viscosity epoxy resin embedding medium for electron microscopy. J. Ultrastruct. Res. **26**, 31—43 (1969).

STUMPF, W. E., ROTH, L. J.: Thin sections cut at temperatures of —70 ° to —90 °C. Nature **205**, 712—713 (1965).

TANDLER, C. J., SOLARI, A. J.: Nucleolar orthophosphate ions. J. Cell Biol. **41**, 91—108 (1969).

TANDLER, C. J., LIBANATI, C. M., SANCHIS, C. A.: The intracellular localization of inorganic cations with potassium pyroantimonate. J. Cell Biol. **45**, 355—366 (1970).

THOMAS, R. S.: Microincineration techniques for electron-microscopic localization of biological minerals. Adv. opt. electron microscopy, R. BARER and V. E. COSSLETT, eds., **3**, 99—154. London-New York: Academic Press 1969.

THOMAS, R. S., CORLETT, M.: The ash-pattern of DNA-phosphate in *Tipula iridescent* virus. In: 3rd int. biophysics congr., Cambridge, Massachusetts, 1969, p. 211.

TOUSIMIS, A. J.: Electron-probe microanalysis of biological specimens. In: 3rd int. symp. x-ray optics microanalysis, Stanford, 1962, H. H. PATTEE, V. E. COSSLETT and A. ENGSTRÖM, eds., p. 539—557. New York: Academic Press 1963a.

TOUSIMIS, A. J.: Electron probe x-ray microanalysis of medical and biological specimens. In: Symp. x-ray and electron probe analysis. Special Techn. Publ. No. **349**, 193—206. Washington: Amer. Soc. for Testing and Materials 1963b.

TOUSIMIS, A. J.: Scanning electron probe microanalysis of biological specimens. Biomed. Sci. Instr. **1**, 249—261 (1963c).

TOUSIMIS, A. J.: Applications of the electron probe x-ray microanalyzer in biology and medicine. Amer. J. Med. Electron. **5**, 15—23 (1966).

TOUSIMIS, A. J.: Electron probe microanalysis of biological structures. In: X-ray and electron probe analysis in biomedical research, A. J. TOUSIMIS and K. M. EARLE, eds., p. 87—103. New York: Plenum Press 1969a.

TOUSIMIS, A. J.: A combined scanning electron microscopy and electron probe microanalysis of biological soft tissues. Proc. 2nd ann. scanning electron microscope symp., Chicago, 1969, O. JOHARI, ed., p. 217—229. Chicago: Illinois Inst. of Technol. Res. Inst. 1969b.

TOUSIMIS, A. J.: Combined scanning electron microscopy and electron probe analysis of biological tissues. Proc. 27th ann. meet. electron microsc. soc. amer. 1969c.

TOUSIMIS, A. J., BROOKS, E. J., BIRKS, L. S.: Possible biological applications of the electron probe. In: X-ray spectrochemical analysis, L. S. BIRKS, ed., p. 122—123. New York: Interscience Publ. 1959.

TOUSIMIS, A. J., MARTON, L., eds.: Electron probe microanalysis. New York-London: Academic Press 1969.

WAISEL, Y., HOFFEN, A., ESHEL, A.: The localization of aluminium in the cortex cells of bean and barley roots by x-ray microanalysis. Physiol. Plant. **23**, 75—79 (1970).

ZIEGLER, H., LÜTTGE, U.: Die Salzdrüsen von *Limonium vulgare*. II. Mitteilung. Die Lokalisierung des Chlorids. Planta **74**, 1—17 (1967).

Note added in proof

Additional references of papers published after the completion of this chapter:

a) Monographs and reviews.

ANDERSEN, C. A., HINTHORNE, J. R.: Ion microprobe mass analyzer. Science **175**, 853 to 860 (1972).

Birks, L. S.: Electron probe microanalysis. 2nd ed. New York: Wiley-Interscience 1971.

Chandler, J. A.: An introduction to analytical electron microscopy. Micron **3**, 85—92 (1972).

Hall, T. A.: The microprobe assay of chemical elements. In: Physical Techniques in Biological Research, 2nd ed., G. Oster, ed., Vol. I, Part A, p. 157—275. New York-London: Academic Press 1971.

Läuchli, A.: Investigation of ion transport in plants by electron probe analysis: principles and perspectives. In: Liverpool Workshop on Ion Transport in Plants, W. P. Anderson, ed. London: Academic Press (in press).

Russ, J. C., ed.: Energy dispersion x-ray analysis: x-ray and electron probe analysis. Special Techn. Publ. No. **485**. Philadelphia: Amer. Soc. for Testing and Materials 1971.

b) Specimen preparation.

Appleton, T. C.: "Dry" ultra-thin frozen sections for electron microscopy and x-ray microanalysis: the cryostat approach. Micron **3**, 101—105 (1972).

Fisher, D. B.: Artifacts in the embedment of water-soluble compounds for light microscopy. Plant Physiol. **49**, 161—165 (1972).

Fisher, D. B., Housley, T. L.: The retention of watersoluble compounds during freeze-substitution and microautoradiography. Plant Physiol. **49**, 166—171 (1972).

c) Microautoradiography and electron probe analysis.

Läuchli, A.: Translocation of inorganic solutes. Ann. Rev. Plant Physiol. **23**, 197—218 (1972).

Index

Bold numbers refer to pages with more extensive treatment, *italic* numbers refer to pages with figures.